A History of Innovation: Pioneering Metathesis for High-Performance Resins

Parkar

Table of Contents

Appendix A: NMR Spectral Data for Selected Compounds

Appendix B: FTIR Spectrum for Selected Compounds

Appendix C: Raman Spectral Data for Selected Compounds

Appendix D: Thermomechanical Testing Results

Appendix E: Other Chemical Structures and Kinetic Study of Polymerization.

Appendix F: Photos of "UVIC" letters

Chapter One: Introduction and Literature Review

1.1.0 History of PDCPD

Robert Minchak, a scientist in Akron, Ohio, from B. F. Goodrich Co., introduced a gel-free copolymer system consisting of cyclopentene (CPE) and dicyclopentadiene (DCPD) with tungsten-based (W) catalysts in 1977. However, this copolymerization required harsh reaction conditions and resulted in unwanted insoluble by-products.[1] Later, in 1983, Daniel Klosiewiez invented an in-mold polymerization method to produce a desirable tough and rigid polydicyclopentadiene (PDCPD) that has high impact strength and high flexural modulus.[2] As shown in Figure 1, a two-part metathesis catalyst system was used to achieve in-mold ring-opening metathesis polymerization (ROMP). The first part of the catalyst system was $WOCl_4$, WCl_6, or a combination of WCl_6 and alcohol or phenol, dissolved in dicyclopentadiene monomer. The second part of the catalyst system contained an activator, specifically, $SnBu_4$, $AlEt_3$, $AlEt_2Cl$, or $AlEtCl_2$, which was likewise dissolved in dicyclopentadiene monomer. In addition, diverse fillers and stabilizers were added to the system to modify the properties of final polymer material. Both mixtures are stable (i.e., do not undergo polymerization) until they are mixed together, at which point the reaction is initiated. During the entire process, water and air should be removed to avoid unwanted hydrolysis or deactivation of both catalysts and activators. Early process optimization made use of solvents (benzene, toluene, chlorobenzene or dichlorobenzene) to help facilitate dissolution of the catalyst. However, subsequent improvements to the reaction injection molding (RIM) protocol led to the polymerization being conducted without solvent. Catalyst loading was well-controlled in the process, with a typical DCPD:tungsten ratio of 1000:1 to 15,000:1 on a molar basis, and a DCPD to alkyl aluminum ratio of 100:1 to 2,000:1. The two streams were injected into a warm RIM mold and then polymerized in a very short time. The product from the RIM process was a rigid solid. Even though metathesis-catalyzed polymerization is inhibited by oxygen, it is not necessary to protect this rapid RIM production under inert gas. Post-polymerization crosslinking took place at 175 °C to remove residual monomer (as well as any solvent that was used), to enhance physical properties, and to finalize dimensional state. The final products typically suffered a 1–3.5 % shrinkage after thermal crosslinking.[2] Afterwards, this RIM technique was acquired by a company called Hercules, where it was modified further. Ultimately, a high-performance thermoset resin which has huge potential in the industry was commercialized and given a new name: Metton®.[2]

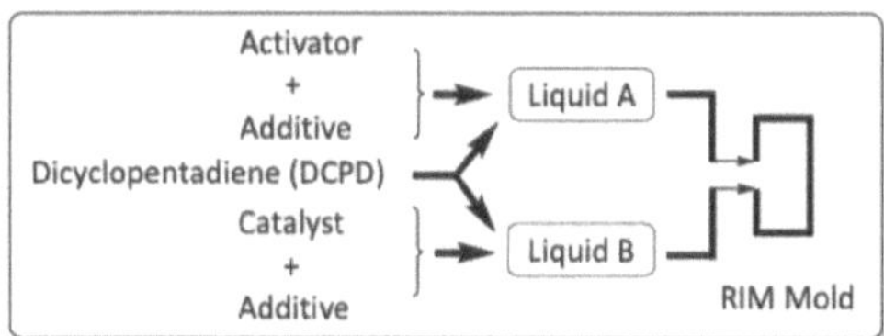

Figure 1. Schematic demonstration of reaction injection molding process (RIM).

More researchers realized the value of PDCPD so that an increasing number of patents have emerged. In the late 1980s, the Hercules company worked on modifying the RIM technique as well as manufacturing PDCPD on a large scale. Afterwards, Hercules researchers Newburg and Tom made critical contributions improving the stability of the tungsten compound/monomer solution and developing cellular PDCPD for use in electrical appliances, electronic cabinetry, furniture, sports equipment, and construction materials.[3-6] Later on, according to a patent that was written by Matlack and coworkers (also from Hercules), copolymerizable cyclopentadiene oligomers were introduced to the DCPD homopolymer system. Giving the fact that the melting point of pure dicyclopentadiene is 32 °C, the addition of oligomers helps to handle dicyclopentadiene in a liquid state at room temperature. The resulting copolymers, which contain 5–30 wt% oligomers, were heated to 125–225 °C for the purpose of removing low molecular weight pyrolysis products. This is one of the earliest demonstrations of PDCPD based copolymerization.[7]

At around the same time, a number of dicyclopentadiene-based copolymers were synthesized that have higher heat distortion temperatures and glass transition temperatures (T_g) than the dicyclopentadiene homopolymers.[8] In addition, PDCPD based matrix polymer composites have been developed. In order to improve stability in air and enhance physical properties upon thermal aging, polymeric particles like thermoplastic polymethylmethacrylate or elastomer polybutadiene were added to PDCPD. The resulting composites have 1 GPa (150,000 psi) notched Izod impact strength at ambient temperatures.[9,10]

Following the developments discussed above, several PDCPD commercial products emerged in the 21st century including Telene® from B. F. Goodrich, Prometa® from Cymetech Co., Metathene® from Hitachi Chemical Co. (under Cymetech's license), Pentam® from Nippon Zeon, Vestenamer® from Evonik Co. Other formulations of PDCPD were marketed by DOW Co. and Shell Co.[11,14,24.]

1.2.0 Ring opening metathesis polymerization (ROMP)
1.2.1 Reaction mechanism of ROMP and living ROMP

Ring-opening metathesis polymerization is a fundamental reaction in polymer chemistry. It has gained more and more attention recently as it is an extremely important method for polyolefin synthesis. Generally speaking, the overall reaction process is defined as the breaking and reforming of double bonds, where a cyclic olefin starting material is converted to a linear olefinic polymer. The main driving force for the polymerization is the release of ring strain.[84]

The basic metal-mediated ring-opening metathesis mechanism consists of 3 fundamental stages: initiation, propagation, and termination (Scheme 1). The reaction is firstly initiated by the coordination of a transition metal-centered carbene complex in a high oxidation state (i.e., a transition metal alkylidene) to a cyclic olefin. Secondly, a four-membered metallacyclobutane intermediate is formed through [2+2]-cycloaddition and this unstable intermediate undergoes cycloreversion (i.e., *retro*-[2+2]) to afford a new metal alkylidene. The overall result of these two steps is to open the strained ring and exchange the carbon fragment that is attached to the metal centre.

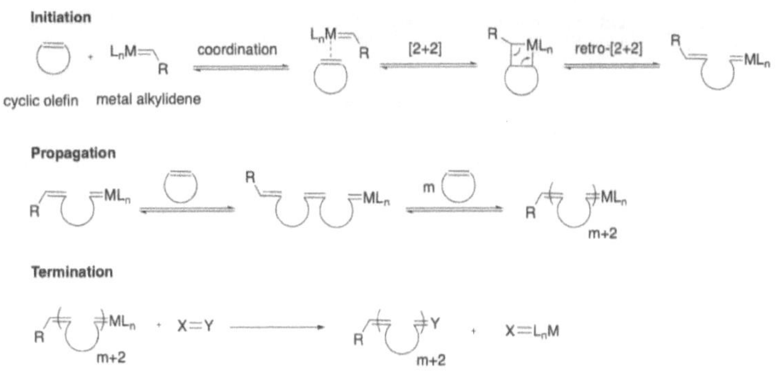

Scheme 1. Basic mechanism of metal catalyzed ring opening metathesis polymerization. Adapted from literature.[85]

The process continues through a series of propagating steps, consuming more monomers until reaching reaction equilibrium or reaction termination. Termination finally takes place when a terminating reagent like ethyl vinyl ether is added to the system to "kill" the active metal alkylidene.[85] Although ROMP is reversible, the equilibrium is controllable thermodynamically. Hence, controlling reaction temperatures and concentrations are critical concerns. In addition, chain transfer is likely to happen through both intermolecular and intramolecular pathways, which are defined as secondary metathesis reactions (Scheme 2). Intermolecular chain-transfer refers to a process in which a polymer chain containing an active metal alkylidene at the chain end reacts with other polymer chains through the backbone. On the other hand, intramolecular chain transfer indicates that the polymer chain reacts with itself, producing oligomers.

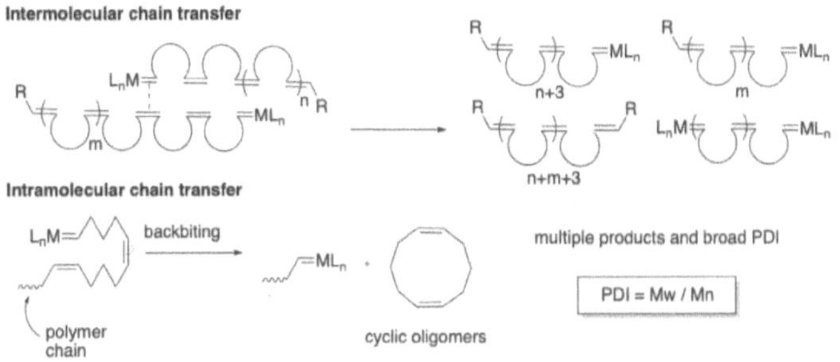

Scheme 2. Typical examples of intermolecular and intramolecular chain transfer. Adapted from literature.[85]

The idea of living polymerization was established by Szwarc in 1956.[84a,b] In a living polymerization system, both premature termination and chain transfer do not take place (or at least are greatly minimized), such that average molecular weight and molecular weight distribution can be controlled by experimental methodology. Additional criteria for living polymerizations include:[85]

i.   A chain-growth polymerization mechanism;
ii.   Fast and complete initiation;
iii. Number-average molecular weight M_n is proportional to monomer conversion;
iv.   Controlled narrow polydispersity (PDI) (typically 1.0–1.1).

When living ROMP meets ruthenium catalysts, it is a powerful tool to design new molecular architectures. For example, novel 3 dimensional branching polymer stars are created with a wide range of functional groups.

1.2.2 ROMP with tungsten and molybdenum

A considerable number of patents and other literature arose in the 1990s, indicating that tungsten and molybdenum (Mo) based catalysts have commercial potential in the manufacture of PDCPD. Polymers made using these catalysts were promoted by Metton (Metton America, Inc.)[15] and Shell (Shell Chemical Company, America, Inc.).[23]

Over the past several decades, there has been a major focus on the development of tungsten and molybdenum based transition metal catalysts for ROMP reactions. The earliest formulation for norbornene polymerization was discovered in the 1950s. This formula used the cocatalyst system of $TiCl_4$ and EtMgBr or $LiAlR_4$ to synthesize polynorbornene.[43-45] In 1986, Schrock developed well-defined new W and Mo carbene (alkylidene) complexes having sterically hindered aryl imido groups and alkoxy ligands that could initiate ROMP in an effective way.[42] Additionally, a molybdenum based complex $Mo(CO)_5Py/C_2H_5AlCl_2/(C_4H_9)_4NCl$ was made for synthesizing polyolefins (polynorbornene and polydicyclopentadiene) through ring-opening metathesis polymerization in 1991. This led to the production of a tough elastomeric solid, but low yields were reported at various temperatures.[43-45]

These catalysts have been modified in order to enhance their stability and functionality. For example, $WCl_6/SnMe_4$ was firstly used as an initiator to produce crosslinked solid PDCPD, while the initiation by $W(OAr)_2Cl_4/SnMe_4$ led to soluble liquid PDCPD. For the purpose of solving the solubility problem, acyclic olefin was added as a transfer agent, which works to lower the molecular weight and reduces network formation in the system.[26] In general, the two-component cocatalyst system is applicable in metal-catalyzed olefin metathesis polymerization. The earlier *in situ* metathesis catalyst systems were unstable and the catalyst reactivity depended on aging time. Those problems were fixed by adding phenoxy ligands to the tungsten or the molybdenum complex and then adding alkyl aluminum chlorides or organotin hydrides as cocatalysts. After that, various new *in situ* catalyst systems were developed with increased activity to shorten induction times during polymerization. Specifically, tin hydride was replaced by transition metal and boron hydrides. In addition, the ligand metal center was changed to new tungsten or molybdenum based complexes in order to increase water tolerance.[23] Recently, a study of a bimetallic catalyst system was carried out by using $(Ph_4P)_2[W_2(\mu\text{-}Br)_3Br_6]$ complexes, which are relatively air stable at room temperature. The catalyst loading was adjusted to 1/250 per DCPD on a molar basis. This formulation resulted in insoluble crosslinked PDCPD.[27] Later on, a study found that the use of a polymeric ligand revealed a significant improvement on polymerization yield and mechanical properties of products. This poly-*p*-acetylstyrene (PAS) macroligand provides enhanced rigidity to the catalyst complex, due to the sterically hindered phenolic groups and aromatic rings.[28a] One of the chief liabilities associated with early W and Mo ROMP catalysts is their lack of stability in air. An important finding showed that aryloxy substituted tungsten precursors combined with organo-tin or organo-lead cocatalysts are air-stable. Using these catalyst systems, both linear and crosslinked PDCPD are achievable with controllable induction time.[28a,b]

Several studies have focused on achieving high stereoregularity and high tacticity in the linear polymer, as well as controlling head-to-tail linkages through the polymerization.[28] There are three different structural factors that affect the tacticity of PDCPD: head-to-head vs. head-to-tail vs. tail-to-tail bonding, *cis* vs. *trans* geometric isomerism in the polymer chain, and *meso* vs. *racemo* sterochemistry of adjacent ring systems in the final polymer product. Further, *isotactic, syndiotactic* and *atactic* hydrogenated PDCPD can be synthesized by the use of several tungsten(VI) imido phenolate complexes.[16] In addition, a new class of molybdenum(VI) and tungsten(VI) complexes were synthesized with formula $Mo(=O)(O\text{-}Ar)_4$, (where $(O\text{-}Ar)_4$ is either four phenolate ligands or two bidentate biphenolates).[19a] Specifically, $MoOCl_4$-biphenolate-*n*-BuLi was developed to control stereo-selectivity of ROMP. As a result, the hydrogenated polymer has a crystalline structure.[19] As shown in Scheme 3A and B, for example, the synthesis of *trans* or *cis* PDCPD was accomplished through the use of tungsten or molybdenum based phenolate alkylidene complexes.[16,19,22] By the same token, research was focused on the development of *syndiotactic* hydrogenated poly(*endo*-dicyclopentadiene) with tungsten complexes that have imido ligands like $W(=N\text{-}R)Cl_4\bullet(Et_2O)_n$ (R=Ph, 2,6-Me_2Ph, -naphthyl, 2,6-*i*-Pr_2Ph, Et, *n*-Bu, *n*-Hex, cyclohexyl, adamantyl; n=0 or 1) or $W(=N\text{-}Ph)Cl_4\bullet Et_2O\text{-}Et_2Al(OEt)$.[20,21] Another similar study was carried out on polymerization to generate *cis, isotactic* linear PDCPD that is catalyzed by $WOCl_4$-based catalyst. It also resulted in a crystalline polymer instead of amorphous material. This stereochemically regular linear PDCPD, which was characterized by wide-angle X-ray diffraction (WAXD) and gel permeation chromatography (GPC), exhibits more than 90% *cis* and 95% *meso* configuration. According to the differential scanning calorimetry (DSC) result, the polymer has thermal stability up to 200 °C.[20] In addition, neutral pentacoordinated and cationic tetracoordinated molybdenum and tungsten imido alkylidene *N*-heterocyclic carbene (NHC) complexes were applied as cationic initiators to prepare highly selective (>98%) *trans-cis* polymer from norbornene derivatives including *endo,exo*-2,3-dicarbomethoxynorborn-5-ene (DCMNBE).[34]

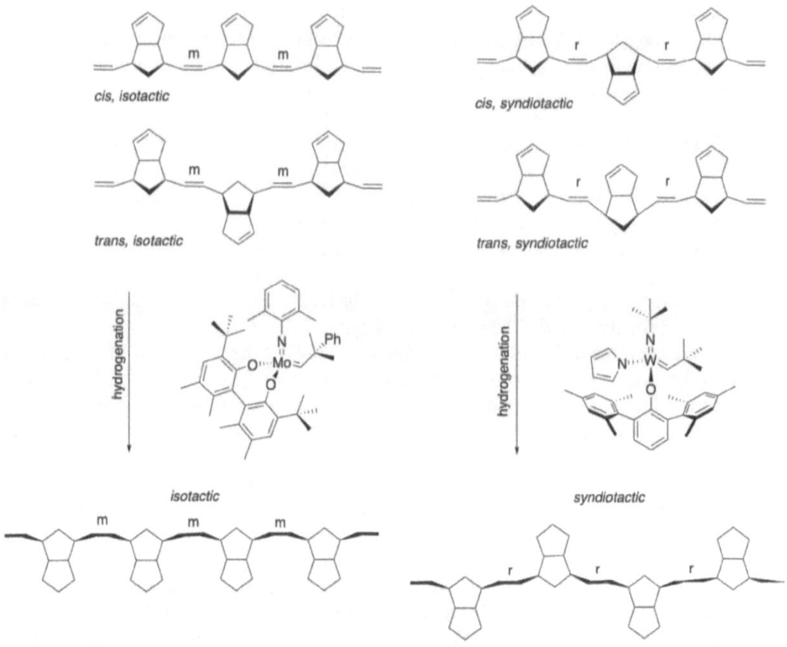

cis, isotactic
cis, syndiotactic
trans, isotactic
trans, syndiotactic
m
m
m
m
r
r
r
r
hydrogenation
hydrogenation
N
Mo
O
O
Ph
N
N
W
O
isotactic
syndiotactic
m
m
m
r
r
r

Scheme 3. A) Stereospecific structures of PDCPD and hydrogenated PDCPD. (m = *meso*, r = *raceme*); B) Selected tungsten and molybdenum complexes that are catalytically active towards the polymerization of *cis,iso*-PDCPD and *cis,syndio*-PDCPD. Adapted from literature.[16,19,22]

Some studies demonstrated the functionalization of W or Mo organometallic complexes for the purpose of increasing reactivity and activity of the catalysts. An important direction in catalyst design is to extend the functional group tolerance. With this intention, high oxidation state Mo imido alkylidene NHC complexes were selected in order to stabilize the cationic charge at molybdenum and facilitate immobilization on silica.[29] Additionally, a series of 2,6-bis[1-(aryl imino)ethyl]pyridine-MoCl$_3$ cocatalysts showed activity towards ROMP of soluble DCPD with high stereoselectivity.[94]

A study illustrated the development of some novel molybdenum imido alkylidene NHC pre-catalysts which bear stronger electron-donating NHC groups. Theoretically, the use of a stronger electron-donating NHC group benefits the polymerization reaction by facilitating the dissociation of an anionic ligand from neutral, high-oxidation state molybdenum imido alkylidene NHC complexes. The PDCPD that was obtained using these modified precatalysts has a wide range of crosslinking degree and glass transition temperatures. In addition, these tailored complexes exhibited dramatically enhanced stability under air.[35]

Several studies have focused on precisely controlling the polymerization to achieve controlled molecular weight and narrow polydispersity. This can often be accomplished by improving the performance of the organometallic catalysts. For instance, a type of bis(catecholato) complex, $WCl_2(catecholato)_2$, was found to be active for ROMP of norbornene and derivatives to provide relatively narrow molecular weight distribution.[18] Also, two bisdiolatotungsten(VI) complexes: $W(diol)_2(phe)_2$ and $W(diol)_2(biphe)$, (where diol=aliphatic diolate dianion, phe=substituted phenolate, biphe=substituted biphenolate) combined with activator Et_2AlCl were studied for ROMP of norbornene and dicyclopentadiene. Among these bisdiolatotungsten(VI) complexes, one binaphtholato complex affords >95% yield after polymerization.[36]

Nevertheless, one significant challenge among tungsten and molybdenum based catalysts is that most of these complexes are air-sensitive, which limits the range of possible applications. One study indicated that WCl_6 was used as a catalyst precursor for producing self-healable materials from *exo*-DCPD. Even though WCl_6 was moderately stable, no healing activity was observed after a certain exposure time under air.[17] More studies focusing on modifying the stability of tungsten catalysts have been carried out over the past 5 years. For instance, in 2018, pentacoordinated 16-valence electron (VE), molybdenum imido alkylidene *N*-heterocyclic carbene complexes, and hexacoordinated 18-valence electron molybdenum imido alkylidene NHC complexes were designed to minimize air sensitivity. Additionally, the fact that these latent precatalysts are thermally stable affords additional benefits for industrial manufacturing. Costs were lowered as no inert gas was needed. By altering the imido and alkoxide ligand, it is possible to tune the glass transition temperature and the swelling propensity of the synthesized PDCPD.[32]

1.2.3 ROMP with Grubbs catalysts

Generally speaking, the properties of transitional metal based catalysts including reaction activity, air stability, and stereoselectivity all depend on the type of alkoxy or NHC ligands, while the functional group tolerance is determined by the metal centre. When moving from left to right across the periodic table, the functional group tolerance increases. Thus, comparing with titanium, molybdenum, and tungsten complexes, ruthenium complexes generally have a better ability to tolerate functionality. In general, appropriately designed ruthenium catalysts can function in the presence of hydroxyl groups, aldehydes, ketals, nitro groups, sulfonates, carboxylic acids, and carboxylic acid derivatives.[85]

Ruthenium-based catalysts are an attractive option for use in ring-opening metathesis polymerization because of their high activity towards olefinic substrates in the presence of most functional groups (Figure 2 shows commonly used and commercially available Grubbs type catalysts). In 1986, $RuCl_3 \cdot H_2O$ was found to effectively initiate the polymerization of *endo*-dicyclopentadiene.[57] The earliest study on $(PR_3)_2(X)_2Ru=CHR^1$ (PR_3=phosphine ligand; X=Cl/Br/I; R^1=Cy/Cp/Ph/Bn), especially $(PCy_3)_2(Cl)_2Ru=CHPh$ (PCy_3=tricyclohexyl phosphine) indicated that the mechanism of polymerization involves phosphine dissociation to free up a vacant coordination site on the metal, followed by Ru coordination to the carbon-carbon double bond (Scheme 4 demonstrates the required association and dissociation of PCy_3 groups). The highest catalytic activity was achieved when the catalysts contained a sterically hindered electron-donating phosphine ligand.[56a,b,c,d] Later on, new ruthenium alkylidenes that bear *N*-heterocyclic carbene (NHC) ligands were discovered. These ligands are larger and are more electron-donating than trialkylphosphines. Replacement of one of the two phosphine ligands on the complex (i.e. to form (NHC)(PCy$_3$)(Cl)$_2$Ru=CHPh) results in increased labilization of the remaining phosphine, which in turn affords a greater concentration of active, coordinatively unsaturated catalyst species and thus a greater rate of polymerization.[55] According to the kinetic data, for a series of L(PR$_3$)(X)$_2$Ru=CHR^1 (where L=L-type ligand; PR$_3$=phosphine ligand; X=Cl/Br/I; R^1=Cy/Cp/Ph/Bn) complexes, L, X, R and R^1 affect the phosphine dissociation rate, olefin metathesis initiation rate, catalyst activity, and alternating ligands effect. Strong evidence was found that higher activity could be achieved *via* various approaches including changing L-type ligands from phosphines to NHCs. Additionally, the initiation rate can be accelerated by switching the halide ligand from chloride to iodide.[55]

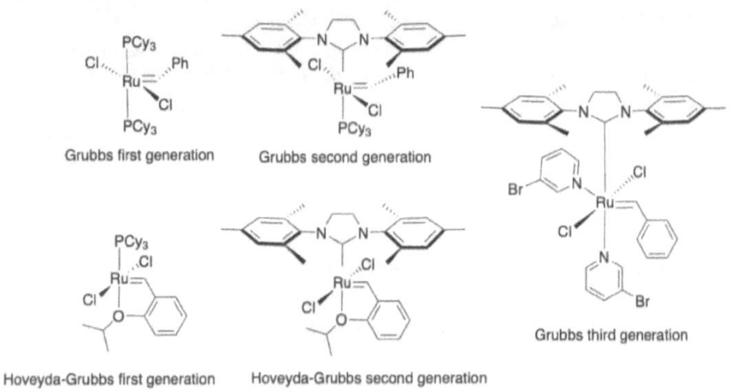

Figure 2. Chemical structures of Grubbs and Hoveyda-Grubbs catalysts.

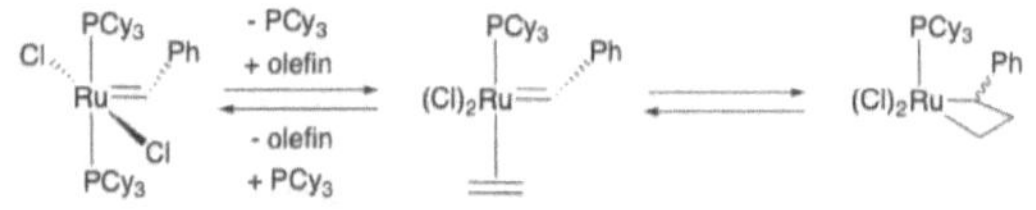

Scheme 4. Mechanism of ruthenium initiation of ROMP. Adapted from literature.[190]

Many studies have focused on the development of novel ruthenium complexes. For example, an aryloxybenzylidene ruthenium (III) complex was described that has potential application in DCPD polymerization.[83] Likewise, a ruthenium complex bearing an isopropoxy-indenylidene bidentate ligand showed dramatically increased thermal stability and high activity in olefin metathesis at elevated temperatures.[168]

Grubbs third generation catalyst (G3) (Figure 2) is often used to afford linear polydicyclopentadiene with well-controlled molecular weight and narrow PDI. In the presence of G3, polymer chains maintain living character with a low catalyst loading.

Conversely, a high catalyst loading results in chain transfer and increased polydipersity. To stop the reaction, two terminating agents are commonly selected: ethyl vinyl ether (EVE) or triphenylphosphine (PPh$_3$). The choice of different terminating agents leads to different thermal properties of the final polymers.[88]

Grubbs and coworkers modified ruthenium-based catalysts to polymerize various sterically hindered or electronically deactivated cyclic olefins. To begin with, replacement of the dimethylvinyl carbene or tricyclopentyl phosphine ligand with a much bulkier ligand for example, tricyclohexyl phosphine (PCy$_3$) helped to speed up the dissociation of phosphine and hence the initiation process was accelerated. At this stage, the catalyst loading was lowered to 100,000/1 on a molar basis. This relatively low loading resulted in a higher polymer yield with shorter reaction times at 50 °C.[79a,b] The modified catalyst can tolerate a broader range of protic and polar functional groups, like alcohols, acids, and aldehydes. Later on, the NHC group was added to the complex, which then showed improved stability at elevated temperatures, as well as better performance under air with increased catalytic activity. Although functionalized, telechelic, and trisubstituted polymers were produced by the improved Ru catalyst, they did not display well-defined polymeric structures or narrow PDI.[79a]

Recent developments in the design of improved ruthenium catalysts have often been focused towards applications like controlling molecular weight and molecular weight distribution.[55] For example, ruthenium based complexes consisting of six-membered nitrogen and sulfur rings were synthesized. The complexes were modified from Grubbs second and third generation catalysts by adding *ortho*-vinylbenzyl-substituted amines or sulfides. These revised catalytic systems exhibited good ROMP activity upon heating.[62] Furthermore, some ruthenium amide complexes, after being activated by trimethylsilyl chloride or HCl, showed high catalytic activity. In addition, a Ru(II) complex that was coordinated by aniline in DMSO, exhibited tunability towards polymerization.[70,71] Meanwhile, new NHC ruthenium complexes like [imidazol(in)-2-ylidene ruthenium] were studied for the polymerization of cyclopentene and norbornene homopolymers.[77] Other novel catalysts were made including a latent non-Grubbs type ruthenium phenylindenylidene complex bearing salicylaldimine ligands,[92] a phosphine-free ruthenium-arene complex,[93] and [RuCl$_2$(phosphine)$_x$(amine)$_y$]-type complexes.[93] Recently, a Ru based heterogeneous catalyst exhibited activity in PDCPD synthesis. After deposition on NaSO$_4$, it can be reused 10 times for hydrogenation.[112]

Several 18-electron ruthenium benzylidene complexes with the formula of (PCy$_3$)[($^kN,O$)-picolinate]$_2$RuCHPh and (H$_2$IMeS) [($^kN,O$)-picolinate]$_2$RuCHPh were synthesized and characterized. Although no ROMP was observed even at elevated temperatures in the presence of these complexes, DCPD was polymerized when the complexes were stimulated in the addition of HCl to some catalyst systems. The use of a strong acid helped convert from 18-electron complexes into highly active 14-electron benzylidenes.[65] Similarly, a ruthenium Schiff base catalyst bearing a NHC type ligand, after being activated by HCl, demonstrated good catalytic reactivity.[73] Also, acid-triggered *N,O*-chelating precatalysts containing 8-quinolinolate ligands showed good catalytic activity towards PDCPD synthesis.[75] Another catalyst system bears substituted pyridine, which is coordinatively bonded to the metal center of a NHC-Ru type complex that contains an alkylidene. Although it is inactive at room temperature, this system can initiate the polymerization of DCPD at 60 °C. Owing to the additional electron donor on the substituted pyridine, it coordinates to the Ru-NHC alkylidene metal center. The additional donor plays a role in raising the initiation temperature without affecting the inherent activity of the catalyst.[81]

One study has focused on exchanging the halide on the ruthenium complexes. A big improvement in the catalytic activity and selectivity was achieved when a chloride was exchanged for an iodide in sulfur-chelated benzylidenes. Also, this latent catalyst system was found effective for norbornene polymerization.[78]

More efforts were made on discovering the best catalyst system for neat polymerization. Cyclic chelating ruthenium phenolate complexes, which are soluble in liquid dicyclopentadiene, were utilized in neat ROMP.[67] Another study was carried out by using molecular sieves as mesoporous materials to immobilize tRuCl$_2$(p-cymene)(PCy$_3$). This new heterogeneous catalytic system was shown to be active in the polymerization of norbornene and its derivatives including dicyclopentadiene.[80]

Plenty of modified Grubbs catalysts were applied in ROMP to pursue different purposes. For instance, the modified Hoveyada-type complexes bear strong coordinating co-ligands that consist of hetero atoms like O, N, S and Se. An extra N-chelating ligand was added to the Hoveyda-Grubbs second generation catalyst (HG2) in order to produce PDCPD in industry at a large scale.[24] A kinetic study of this catalyst was carried out in 2019 which determined the relationship of activation energy and degree of polymerization. It also suggested that the crosslinking was taking place through the double bonds on cyclopentene rings of DCPD (see below for further discussion about the mechanism of PDCPD crosslinking).[82] Additionally, ionic Grubbs-Hoveyda complexes were used for both homogeneous and biphasic liquid-liquid polymerization. These complexes showed catalytic activity towards cis-cyclooctene and dicyclopentadiene homopolymerization with good yield and relatively low metal content between 10–80 ppm.[66] Another study indicated that M2 and M22, two less common but commercially available ruthenium-based initiators, also exhibited activity to produce PDCPD in high quality.[61]

Recent studies in the polymerization of DCPD demonstrated that both *endo* and *exo* dicyclopentadiene produced polymers are insoluble in organic solvents, which indicates the formation of crosslinked polyolefins.[33] According to a kinetic study, the ROMP reactivity of the *exo*-dicyclopentadiene is proved to be higher than the *endo*-dicyclopentadiene in the presence of the Grubbs catalyst.[91]

1.2.4 Other metal-based catalyst systems

Other transition metals like tantalum (Ta), vanadium (V), and titanium (Ti) have also been used in the development of metal-based catalysts for ROMP. For example, in 1987, Schrock and Wallace made some contributions to developing a tantalum alkylidene complex Ta(CH^{t}Bu)(OR)$_3$ for the synthesis of polynorbornene (PNEB).[40] Similarly, another tantalum metal centered alkylidene complex containing the mono(*ortho*)-chelating diamine ligand (CNN), showed good reactivity in the production of PNEB.[30] Researchers also designed (arylimido)vanadium(V)-alkylidene catalysts for norbornene (NEB) and tetracyclododecene copolymerization. As a result, the reaction affords high molecular weight polymers (M_n=19,500–47,400) with low polydispersity index (PDI=1.18–1.38). Also, the hydrogenated products have melting points (T_m) around 294 °C.[31] One interesting experiment has demonstrated that a nanoporous polydicyclopentadiene thimble was used as a matrix to study the retention effect of palladium and phosphine ligands. The thimble had Pd residues left from polymerization. After being isolated and purified in a high yield, it had no detectable concentration of Pd or phosphine.[60]

Since Karl Ziegler and Giulio Natta were awarded Nobel prize in 1963, the Ziegler-Natta catalysts containing titanium complexes were used for ethylene and propylene polymerization. Afterwards, in the 1980s, Grubbs and his coworkers modified a titanium carbene precursor "Tebbe Reagent" and titanacyclobutanes as ROMP catalysts.[45-47] Further, a cocatalyst system consisting of titanium tetrachloride adduct complexes bearing nitrogen or oxygen containing ligands and methyl lithium $TiCl_4 \cdot 2L/CH_3Li$ (L=pyridine, 2-methylpyridine, 2,4,6-tri-methylpyridine, 3-aminopyridine, 2-hydroxypyridine, dioxane, 2,5-dimethylfuran, or tetrahydrofurfyl alcohol) showed excellent activity in ROMP. In addition, the aging time and aging temperature were controlled by altering the pyridine derivatives. However, these Ti based cocatalysts had no obvious effect on molecular weight distribution.[37,38] Furthermore, Cp_2TiCl_2 (Cp=^{5}-cyclopentadienyl) coupled with Grignard reagents RMgX (R=CH_3, C_2H_5, i-C_3H_7, n-C_4H_9, n-C_6H_{13}, C_6H_5; X=Cl, Br, I) exhibited medium reactivity to the homopolymerization of norbornene and DCPD.[39] One study on titanium complexes like dimethyltianocene, cyclopentadienyltrimethyltitanium(IV), chlorodimethylcyclopentadienyltitanium(IV), and bis[(trimethylsilyl)methyl]titanocene showed the catalytic activity of these complexes in the polynorbornene synthesis. Notably, the use of a polar solvent, for example, tetrahydrofuran (THF), inhibited the polymer reactivity.[46] Later on, this Ti-based catalyst system was modified to dimethyl titanium dichloride. The highest catalyst activity was presented in the presence of Me_2TiCl_2 and polymerization was completed in 5 minutes at 40 °C.[41] Similarly, dicyclopentadiene was polymerized by adding dimethyltitanocene to the solution. At 70–90 °C, linear PDCPD was produced, but when the temperature was above 90 °C, crosslinking occurred.[48] Another modification of Ti-based catalyst was to establish a multicomponent catalyst system which contained bis(cyclopentadienyl)titanium dichloride and diethylaluminum chloride. This catalyst showed activity in the polymerization of DCPD when toluene was used as a solvent.[49]

1.2.5 Photoinitiated and metal-free ROMP

Metal catalysts enable the production of both linear and crosslinked PDCPD. Early developed transition metal photocatalysts like [bis(arene)Ru(II)] exhibited photochemical activity in the initiation of ROMP under UV light. ROMP that is photochemically initiated is defined as photo-induced ring-opening metathesis polymerization (PROMP).[89]

Several novel, photochemically active transition metal catalysts were developed to manufacture PDCPD for both experimental and industrial purposes. For example, new photoswitchable nonchelated ruthenium benzylidene complexes were designed. These catalysts contain benzylphosphite ligands. As an illustration, upon radiation of visible light, these photoinitiators exhibit photoactivity for preparing multi-layered PDCPD films, which are potentially used in 3-D printing applications.[63] Also, surface-initiated polydicyclopentadiene films were produced *via* surface-initiated ring-opening metathesis polymerization (SI-ROMP) with less than one minute on gold and silicon substrates. These 400 nm thickened films were grown both in the vapor phase and in the solution phase. Thin films produced in this way showed relatively high stiffness.[64] In addition, a novel Ru-alkylidene complex that bears a triazene-based chelating carbene ligand was used as a precatalyst for UV-induced polymerization of cyclic olefins including DCPD.[69] In 2019, the most recent study was focused on the photoacid generator-initiated ROMP of cyclooctadiene (COD). This photoinitiated ROMP was carried out with second generation Grubbs catalyst (G2) and modified G2 (SG2 and AG2) under UV light. By using this method, the produced PDCPD resin exhibited a storage modulus (G') of up to 100 MPa.[50]

The metal residues that are trapped in polymer chains has become a potential problem for some specific applications. For instance, these toxic residues limit the use of ROMP materials for the applications in the human body. To solve this problem, PROMP was recently accomplished using a photoinitiated metal-free reaction (Scheme 5). The reaction is initiated by the single-electron oxidization of a vinyl ether or enol ether initiator **A** in the presence of visible light. The excited radical cation **B** is therefore produced through the oxidation procedure. When cycloolefin is involved in the reaction, [2+2] complex **C** is formed. The unstable four membered-ring contained in **C** then rearranges to afford intermediate **D**. Propagation starts since more cycloolefins take part in the reaction and chains keep growing to yield a polymer **E**, which bears an alive radical cation at the chain end. This reversible oxidation-reduction cycle continues as a termination-reinitiation cycle until polymerization is terminated by removing the light to produce **F**. Thiol-ene reactions were then used to crosslink the linear PDCPD product.[51-53] In addition, with the photoredox-mediated metal-free technique, the tacticity of poly(*endo*-dicyclopentadiene) in living polymerization is controllable. In that case, it was estimated 55 % *syndiotactic* configuration among PDCPD chains.[54]

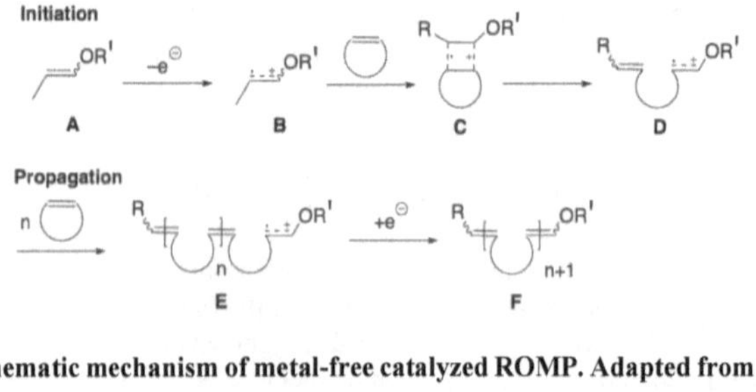

Scheme 5. Schematic mechanism of metal-free catalyzed ROMP. Adapted from literature.[51,52]

1.2.6 Crosslinking (curing) mechanism of PDCPD

The exact mechanism of PDCPD crosslinking depends on the conditions and catalysts used for the polymerization mechanism, and can sometimes be a matter of debate. Most authors have hypothesized that the predominant mode of crosslinking is through secondary ring-opening metathesis of the pendent cyclopentene double bonds within the linear telechelic polymer (Scheme 6).[104] This assumption had not been disproved for 25 years until Wagener showed evidence for thermally crosslinked PDCPD in the presence of tungsten-based catalysts. Wagener's research showed that—at least for early transition metal catalysts—the crosslink does not arise from secondary metathesis events, but rather through olefin addition processes (Scheme 6).[105] Afterwards, more studies were focused on the kinetics of crosslinking with various ruthenium-based complexes. The newer catalytic systems like Grubbs catalysts, Grubbs-Hoveyda catalysts and modified ruthenium catalysts (for example, $RuCl_2(H_2IMeS)$) are more active for the ring-opening metathesis of unstrained cyclopentenes, and can therefore lead to metathesis-type crosslinks. The most compelling structural evidence for this type of crosslink comes from a recent study by the Johnson group, where cleavable linkers placed within the PDCPD backbone allowed for the isolation of soluble, crosslinked molecular fragments that could be fully characterized.[187a] Ruthenium catalysts can also lead to a back-biting mechanism (Scheme 7) due to their high catalytic activity.[106b,c] Of course, the presence of ruthenium catalysts do not preclude the formation of thermally initiated olefin addition-type crosslinks, especially if high temperatures are reached during the exothermic polymerization reaction.

Additionally, different types of catalysts can have an impact on the crosslinking density and physical properties of PDCPD. For example, PDCPD that was catalyzed by G1 has higher crosslinking density but lower tensile strength then the G2 produced polymer.[106a,d] Other factors that affect the crosslinking include catalyst loading and crosslinking temperature.[107] Further study of crosslinking kinetics demonstrated that the activation energy for crosslinking was lowered when using Ru based catalysts rather than W and Mo centered complexes.[24]

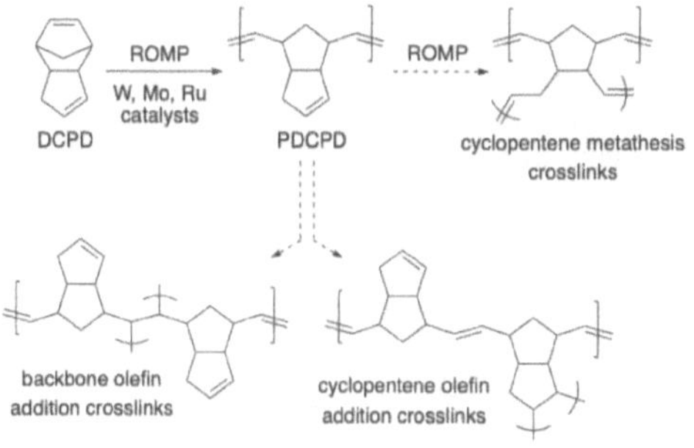

Scheme 6. Proposed mechanisms of crosslinking of PDCPD. Adapted from literature.[191]

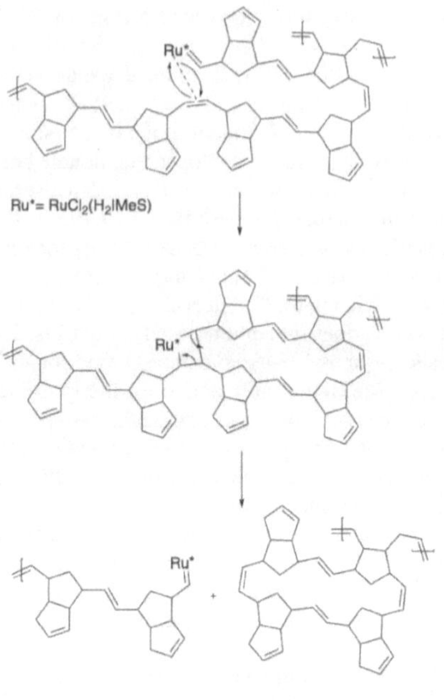

Scheme 7. Backbiting crosslinking mechanism. Adapted from literature.[106a,b,c,d]

In order to investigate the crosslinking behavior of PDCPD, various characterization methods have been used, including differential scanning calorimetry (DSC), dynamic mechanical analysis (DMA), Raman spectroscopy, Fourier-transform infrared spectroscopy (FTIR), liquid and solid state nuclear magnetic resonance (NMR), scanning electron microscopy (SEM), and transmission electron microscopy (TEM). Further studies on the rheological and mechanical behaviour of crosslinked PDCPD have also been carried out for industrial purposes.[108a,b,c] Owing to its crosslinking mechanism, PDCPD can also be used as to enhance the network property of epoxy resin.[109b] Further applications can include self-healing materials, PDCPD based matrices,[109d] binders,[109a] and glass fiber reinforced thermosets.[109c]

Several studies were carried out to design crosslinked copolymer networks. For example, crosslinking agents like norbornene derivatives were used to improve thermal properties without sacrificing mechanical performance.[110a,b,c] Also, crosslinked polymer blends of poly(*endo*-dicyclopentadiene) and poly(5-ethylidene-2-norbornene) were studied kinetically.[110a] The result indicated a decrease in glass transition temperature and crosslinking density as an increasing amount of ENB was added.[110d] Additionally, polybutadiene, an elastomer that can be used as a crosslinking agent, was incorporated as pendent chains or fixed segments in the PDCPD crosslinked network. The elastomer improved flexibility relative to the parent material.[110d] Another strategy to reduce brittleness is to control the molecular weight between crosslinked sites of the PDCPD homopolymer as well as the norbornene-based copolymers.[110e] Consequently, these approaches can engineer polymers at a molecular level to improve the designed physical properties and mechanical performance.

Notably, a molecular simulation study was carried out to investigate the overall behavior of PDCPD upon thermal crosslinking. Functions were built up to illustrate the relationship between degree of crosslinking, glass transition, and thermal expansion coefficient.[113a] In addition, the thermodynamic behaviors of linear PDCPD were assigned to each peak on a DSC curve. The endothermic process is the glass transition, the first exothermic peak is associated with crosslinking through backbone double bonds and the second exothermic peak indicates the thermo-oxidation of linear PDCPD.[113b,c] Also, glass transition temperature, degree of crosslinking and crosslinking density values altogether were determined by DSC using PDCPD produced from different catalyst systems.[113b,c]

It has been conclusively shown that highly crosslinked PDCPD has excellent ballistic performance. The penetration resistance of crosslinked PDCPD was compared with an epoxy/diamine system. Surprisingly, the PDCPD exhibited 300–400 % expanded ballistic energy dissipation in a wider temperature range than the epoxy/diamine composite.[153]

Aging, in another word, oxidation, limits the application of materials in aggressive conditions. It causes degradation and decomposition which lead to the color changing, mechanical properties reducing, and other problems.[111a] Oxidation takes place as long as the material is exposed to air for a certain time at room temperature or upon heating. Also, thermal oxidation was observed on the surface of PDCPD/glass fiber composites that results in surface damage on the interface of composites.[111c] According to many kinetic studies, residual organometallic catalyst accelerates the oxidation on account of favored hydroperoxide (which formed and consumed during the PDCPD oxidation) decomposition into radicals.[111b] FTIR and NMR data indicated that the oxidation results in the consumption of double bonds and the concomitant formation of carbonyls or aldehydes. In fact, the complicated oxidation process is affected by oxygen pressure, temperature, catalyst types, catalyst loading, sample thickness, and characterization methods.[111d,e,f] When the temperature is below the glass transition temperature of the PDCPD (which means that the polymer is in its glassy state) PDCPD oxidizes faster than polybutadiene and polyisoprene under the same conditions.[111e]

1.2.7 Copolymerization

Schrock catalysts and Grubbs catalysts can not only be applied in homopolymerization but also in copolymerization. Recently, there has been renewed interest in the copolymerization of norbornene and its derivatives. For example, copolymerization of cyclopentene with 5 % strained olefin (norbornene) was studied in a purified ruthenium-catalyzed system. The maximum conversion of cyclopentene was 60 %.[58] Also, a new class of thermoset composites was synthesized that consists of organic components like dicyclopentadiene together with the inorganic norbornenylethyl polyhedral oligomeric silsesquioxane (1NB-POSS) and tris(norbornenylethyl)-POSS (3NB-POSS). The thermosets exhibited flexibility and displayed reduced crosslinking density, compared to crosslinked PDCPD homopolymers.[90] Additionally, the copolymerization of dicyclopentadiene and norbornene derivative 5-norbornene-2,3-dicarboxylic acid methyl ester (NDM) was carried out in the presence of G2. The resulting copolymers were characterized by gel permeation chromatography (GPC) and DSC. These materials had good thermal stability and narrow polydispersity (PDI=1.4).[76] Lately, various telechelic copolycycloolefins consisting of DCPD and cyclooctene were synthesized through ruthenium-catalyzed chain-transfer ROMP. These polyolefins remained liquid at room temperature and exhibited relatively low viscosity.[99]

Several studies were carried out on developing latent catalyst systems for synthesizing norbornene based copolymers. For example, exo-1,4,4a,9,9a,10-hexahydro-9,10(10,20)-benzeno-1,4-methanoanthracene (HBMN) and Pd-N-heterocyclic carbene complexes showed catalytic activity in copolymerization.[87] Additionally, 5-n-hexylnorbornene that was activated by methylaluminoxane (MAO), borate $\{Na^+[B(3,5-(CF_3)_2C_6H_5)_4]^-$ (NaBARF)\} or palladium based catalysts were applicable for copolymerizing norbornene derivatives.[25a] The resulting polymers have relatively high molecular weight and are obtained in good yield. In addition, these products exhibited enhanced thermal stability with porous structures so that they can be used as fillers in composite production for gas storage purposes. Notably, hydrogeneration was also carried out to avoid oxidation of residual double bonds, and to improve stability.[25b] Also, the complex cis,fac-[RuCl$_2$(DMSO-O)(DMSO-S)$_3$ that was preactivated with aniline was utilized as a catalyst for the preparation on PENB/PDCPD or polynorbornadiene/PDCPD. The highest yield was 42 % when the molar ratio of DCPD/ENB was 20/1.[71] What is more, copolymers of PDCPD/polycyclooctene (PCOE) were polymerized under acidic aqueous microemulsion conditions. The reaction was catalyzed by ruthenium alkylidene complexes bearing pH-responsive N-hetereocyclic carbene and N-donor ligands at elevated temperatures. The conversion reached 100 % when the DCPD/COE was about 50 % on a molar basis.[72] Additionally, a catalytic system of W(=NPh)(2,6-Me$_2$-phenolate)$_4$/n-BuLi was applied to polymerize comonomer mixtures containing dicyclopentadiene and tricyclopentadiene (TCPD). Hydrogenation of copolymers was followed up in the presence of a Ru based complex like (PCy$_3$)$_2$Cl$_2$Ru(H)(Cl)(CO). Additionally, a nickel-based complex nickel acetylacetonate Ni(acac)$_2$ was developed by Zeon Co. When the complex is activated by triisobuytlaminum Al(iBu)$_3$, it can also be used to hydrogenate PDCPD.[74]

DCPD can also react with other alkenes. For example, polymerizations of dicyclopentadiene that was mixed with 50 mol% 1,4,5,8-dimethano-1,2,3,4,4a,5,8,8a-octahydro-naphthalene (DMON) were conducted at room temperature in the presence of WCl$_6$/iBu$_3$Al/2-BuOH catalysts. The produced copolymers were then hydrogenated by Ni(acac)$_2$/iBu$_3$Al at 80 °C. As a consequence, the resulting copolymers exhibited improved thermal stability, oxygen resistance and a glass transition temperature of 40 °C. The copolymers had high average molecular weight (M$_w$>4×10^4) and the PDI value was larger than 2.4.[86] Additionally, a kinetic study was carried out on the copolymerization of DCPD and tricyclo[5.2.1.0]dec-8-ene (TCD) that was catalyzed by bis(tricyclohexylphosphine)benzylidene ruthenium dichloride using ultrasonic spectroscopy. The result indicated that by adding TCD (which lacks a second double bond to participate in crosslinking reactions), the crosslinking density of the copolymers was decreased, although the acoustic modulus reached 3.5 GPa.[103]

Several studies have described the design of advanced copolymer architectures using living ROMP with Grubbs catalysts. For instance, a multi-arm star polystyrene (PS) was functionalized with Grubbs catalyzed polyolefins through alkene double bonds in the PS core.[59] Additionally, a block copolymer polylactide-*b*-polynorbornenylethylstyrene was designed as a structural template for the polymerization of DCPD; after polymerization, the polylactide was etched to afford nanoporous structures. This well-structured block copolymer has self-assembled bicontinuous phases, and thus it may have further applications as gas/liquid transports and ultrafiltration membranes.[152]

The most recent studies on photoinitiated ROMP (PROMP) likewise identified the feasibility of copolymerization. As an illustration, a NHC type photogenerator that was combined with an inactive metathesis catalyst was applied to photoinduce norbornene and dicyclopentadiene copolymer films under UV irradiation. Notably, this method enables the crosslinking of copolymer films by only one step.[68] Also, a metal-free copolymerization was developed to synthesize a series of PDCPD based random copolymers with higher molecular weight than was obtained for homo polydicyclopentadiene.[52] Di-block copolymers were synthesized by metal-free organocatalyzed ring-opening metathesis polymerization with bifunctional macroinitiators and a one-pot sequential monomer addition.[53]

1.2.8 Frontal ring opening metathesis polymerization (FROMP)

Frontal polymerization (FP), discovered by Russian scientists Chechilo and Enikolopyan in 1972,[115] is a process that converts monomers into polymers through a localized exothermic reaction that takes place in a moving region of the reaction cell. Propagation occurs within the region *via* the coupling of Arrhenius reaction kinetics and thermal diffusion.[116] In order to initiate a FP, the polymerization must be an exothermic and rapid reaction at adiabatic conditions. In addition, pot life should be long enough, in other words, the reaction must proceed slowly at ambient temperature.[116] Pot life is defined as the point where FP no longer continues because of reagent decomposition or reaction. It can be quantified by measuring the the time required for the mixture to reach the gel point under isothermal conditions. In a rheological measurement the gel point is determined as the crossover of shear storage modulus and shear loss modulus.[122b] As shown in Figure 3, the rapid FP process transforms a liquid or gel mixture of monomers, catalysts, and inhibitors into solidified polymers. Fronts start upon thermal activation and produce heat for further propagation.

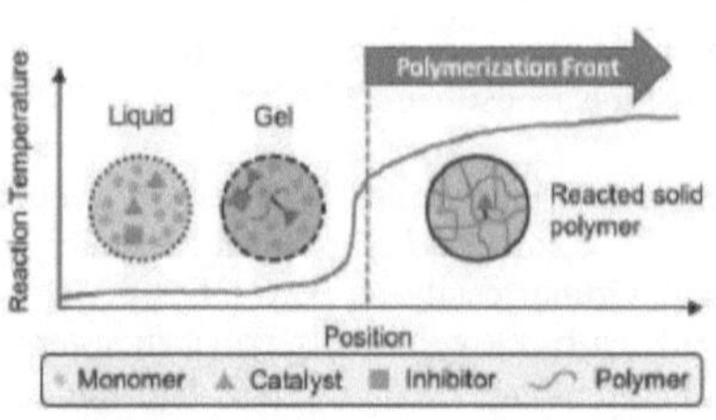

Figure 3. Concept of frontal polymerization and its application toward manufacture of PDCPD based composites. Adapted from literature.[122]

The first example of dicyclopentadiene frontal ROMP (FROMP) was accomplished in 2001, from which a fundamental set of successful reaction conditions was determined. In this method, solid DCPD was mixed with Grubbs catalysts and inhibitor triphenylphosphine (PPh$_3$) at room temperature in a glass test tube. The polymerization was then initiated by heating the wall of the reaction vessel with a copper-Constantan thermocouple. Experiments were performed in the descending mode (i.e. with the front propagating down) to avoid buoyancy convection. Additionally, to store the mixture of inhibited catalyst and monomer at 7–8 °C for 3 weeks, more PPh$_3$ was added to balance the competing demands of a long pot life and a rapid reaction.[115] This ligand strongly binds with the ruthenium metal center of Grubbs catalysts at room temperature and therefore deactivates the catalyst. As the temperature increases, PPh$_3$ dissociates from the metal center so that the catalyst is reactivated for engagement with DCPD. Also, the entire inhibition/activation process is reversible.

Several studies have been carried out in the development of novel inhibitors for FROMP. For instance, N-donor ligands like 1-methylimidazole, 4-(N,N,dimethylamino)pyridine (DMAP), and pyridine acted as inhibitors in the polymerization. The inhibition effect lasted up to 72 hours, while the inhibited system was reinitiated rapidly in the addition of H$_3$PO$_4$.[117,118] Another study showed that when DMAP was mixed with GC2 and DCPD at 1:1:2,000 on a molar basis, frontal velocity reached 15 cm/min which was the highest value reported, and the maximum frontal initiation temperature (T_{max}) was 205 °C.[120] Limonene, a renewable resource that is commonly used as a solvent or a chain transfer agent in ROMP, was applied in FROMP.[119] Additionally, limonene was also copolymerized with DCPD resulting in copolymers which have increased pot life. These copolymers have a high swelling ratio and good mechanical properties. Limonene also acts as a plasticizer and a chain transfer agent to control the crosslinking density of produced copolymers. However, the stiffness of the produced homopolymer and copolymer was decreased.[119] Afterwards, a kinetic study of FROMP was carried out to observe heat transfer and heat effects of thermal triggered polymerization of DCPD. It illustrated that uneven heat was dissipated during polymerization, and the heat transfer depends on the convection of the system, according to the high-resolution infrared (IR) images.[116]

Several systematic studies have been carried out by the Moore group, including Grubbs catalysts loading, improving frontal velocity, and altering inhibitors type. They pointed out that exo-DCPD undergoes a higher FROMP rate than endo-DCPD.[122a] Also, endo-DCPD cracks to cyclopentadiene quickly at high temperatures and that introduces unwanted voids to the final crosslinked PDCPD samples. When DMAP was used as the inhibitor (at an optimized ratio of 16:1 vs. Grubbs second generation catalyst) and 100,000:1 catalyst:substrate loading, the pot life, however, was reduced to 10 minutes. According to the kinetic study result, it is difficult to balance the pot life and frontal velocity as they are strongly inverse to each other. Notably, a later study demonstrated an improved formula based upon switching the inhibitor from DMAP to an alkyl phosphite (tributyl phosphite).[122b] The alkyl phosphite is presumed to bind strongly to the metal center of the ruthenium alkylidene at room temperature. With this formula, hence, the pot life was improved to 30 hours with an even lower catalyst loading.[122b] At this stage, the methodology was well established and it allowed the complete control of FROMP for further applications. Notably, in early 2018, a significant work on rapid energy-efficient manufacturing 3-D printable DCPD resin *via* FROMP attracted much more attention in this field (Figure 4). In this work it was conclusively shown that FROMP is an efficient method to rapidly polymerize and fully crosslink polymer within seconds at a very low level of energy consumption. This method was applied to the fabrication of carbon fiber reinforced PDCPD composites and 3-D printed structures with microscale features. Compared to aero-space grade epoxy, the DCPD resin exhibited a similar high performance.[122c] Consequently, these findings lead to the potential application of FROMP-produced PDCPD in the field of automotive engineering as well as the aerospace, marine and energy industries. This novel technique also has potential usage in moulding, imprinting, and 3-D printing techniques.[122c]

Copolymers of norbornene and its derivatives can also be produced *via* FROMP. For instance, dicyclopentadiene and methyl methacrylate or tri(ethylene glycol) copolymers were produced in the presence of Grubbs catalysts and a PPh_3 inhibitor.[121] The copolymer system containing DCPD and dinorbornenyl (di-NBE) crosslinker derivatives is essentially able to tune the material properties, for example glass transition temperature.[122c,d] The most recent research was focused on the frontal copolymerization of dicyclopentadiene and 1,5, cyclooctadiene (COD). The glass transition temperature of these poly(dicyclopentadiene)-*ran*-poly(1,4-butadiene) materials is tunable in a range of −90 to +114 °C along with tunable tensile moduli from 3.1 MPa to 1.9 GPa. Hence, the addition of an elastomer enables the rapid manufacture of elastomeric materials for shape memory purposes.[189]

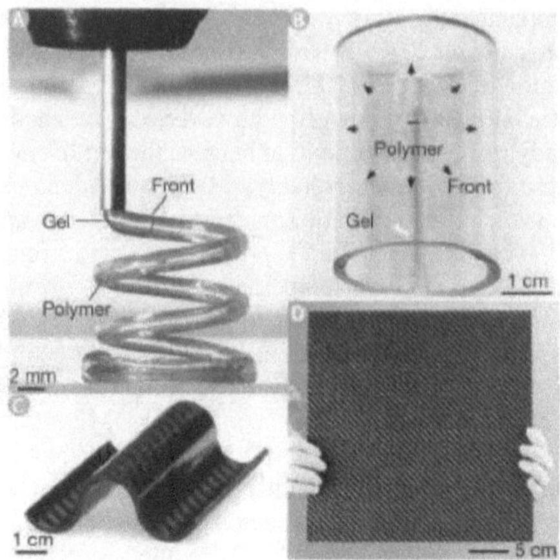

Figure 4. Products of advanced manufactured PDCPD with FROMP. A) 3D printing of gel DCPD solution that is solidified by FROMP immediately following extrusion from the print head; B) a free standing gel, propagating radially from a single initiation-point source; C) a corrugated carbon composite fabricated by FROMP; D) A carbon composite panel cured by FROMP in 5 minutes. Adapted from literature.[122]

One study on simulating the manufacturing process of unidirectional PDCPD fiber composites *via* frontal polymerization indicated that the fiber volume fractions played an important role in controlling the frontal velocity.[188,304] This study helped to establish the relationship between experimental FROMP and simulated data in many aspects like frontal velocity, maximum temperature, and curing kinetics parameters.[188,304]

1.3.0 PDCPD processing method – reaction injection molding (RIM)

Reaction injection molding (RIM) processing is a technique to manufacture thermoset polymers and it has been globally applied in industry.[185,186a,b,c] During a typical PDCPD RIM process, for example, Metton® was produced through a liquid molding resin (LMR) system. In this system, one stream contained DCPD, tungsten catalysts (WCl_6 and $WOCl_4$), nonylphenol (solvent to dissolve catalyst), additives like antioxidants, and fillers. The other stream had DCPD, a cocatalyst ($EtAlCl_2$), retarders, other additives, and fillers. Polymerization began with the mixing of two streams in the mold where short induction time was controlled. Additionally, to produce Telene®, precatalysts that consist of tetrakis(tridodecylammonium)octa-molybdate, an activator containing Et_2AlCl, a trimer of cyclopentadiene, along with propanol and $SiCl_4$ altogether acted as crosslinkers during polymerization.[125] Later on, Materia Co. improved this technique with well-defined Grubbs catalysts to reduce odor from monomer residues.[125]

A systematic study of PDCPD RIM production has been carried out. As a result, a regulated protocol was finalized for the liquid molding system of Telene®. In general, chemical reactions that take place in RIM follow the following fundamental rules:[126b,c]

i)   Components are dissolved and mixed at an appropriate viscosity;
ii)   The reaction is fast enough (usually complete within several minutes);
iii)   Rapid crosslinking or phase separation results in solidification in the mold.

Within the system, gelation time was adjusted to occur on a timescale of minutes. As a consequence, the RIM-produced PDCPD had a tensile modulus of 1860 MPa, and a glass transition temperature at 150 °C. PDCPD produced in this way also possessed low moisture absorption, good chemical resistance, and good paint adhesion. Moreover, a copolymerization of norbornene derivatives has been studied in the same RIM system. It was shown that the calculated autocatalytic kinetic model could predict the reaction kinetics very well. Measured adiabatic rheokinetic results from experiments fit well in this model.[126a] This model, therefore, was applied to predict parameters like the practical rate of PDCPD formation and viscosity during the RIM process.[126a] In addition, the factors that affect dry friction behavior of thermosets depend on the initiation temperature, quality of mixture and sliding velocity.[126d]

Several studies have demonstrated an ability to conduct copolymerization in a RIM mold. For example, copolymers of DCPD/cyclopentene (CPE) (Scheme 8) that were catalyzed by Grubbs catalysts exhibited lower weight loss rate than the corresponding homopolymers. Comonomers were fully converted to copolymers. The copolymers also gained improved mechanical properties. As a consequence, by manipulating the catalyst loading and reaction temperature, it is feasible to alter the gelation time and glass transition temperature.[127]

Scheme 8. Copolymerization and crosslinking of dicyclopentadiene and cyclopentene. Adapted from literature.[127]

RIM is also a powerful technique to manufacture PDCPD based composites. In one illustrative example, DCPD was firstly mixed with polyethylene wax (PEW) and Schrock catalysts. The mixture was then injected into a simulated RIM mold under nitrogen to form PDCPD/PE composites. The bending strength and hardness of the composites were enhanced significantly relative to the parent PDCPD as well as wear resistance. But the friction coefficients were reduced. As a result, mechanical properties of these composites were enhanced with relatively low polyethylene content (less than 3 wt%).[128a,b] Moreover, this simulated RIM process was also applied in other PDCPD composites materials such as expanded graphite nanosheets which were treated with silane coupling agents (Figure 5E). In addition, mechanical and tribological tests were undertaken in dry sliding and oil-lubricated conditions, upon which the composites showed a lower wear rate and increased friction coefficients.[130]

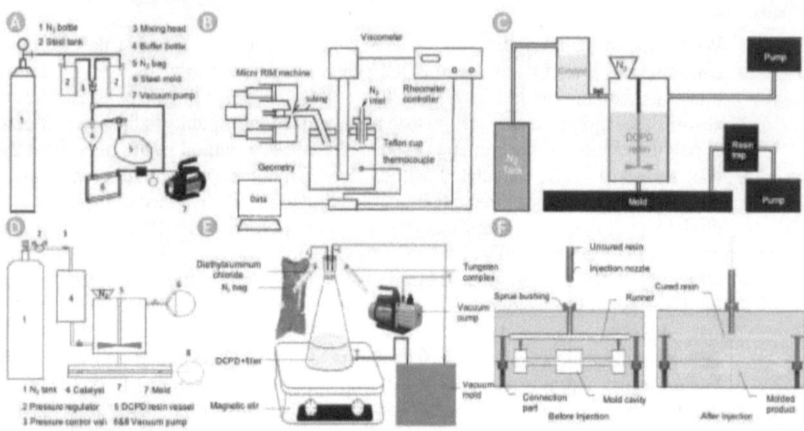

Figure 5. Schematic demonstration of A) a simulated RIM device for the purpose of preparing SiO₂/poly-DCPD nanocomposites; B) a simulated RIM system coupled with rheometer for copolymerization at adiabatic temperature; C) specially designed and customized S-RIM equipment for the purpose of minimizing voids formation; D) RIM process at relatively low pressure; E) simulated RIM system for processing treated expanded graphite nanosheets/polydicyclopentadiene (TEG/PDCPD) composites; F) structural reaction injection molding (S-RIM) process. Adapted from literature. [109a,b,c,d,126a,b,c,d,129,130,146,163]

Several studies have demonstrated lab-scale structural reaction injection molding (S-RIM) devices (Figure 5C, D and F). These devices were modified from industrial RIM setups. They consist of a steel mold, a mixing head, a vacuum pump and nitrogen gas flow.[163] Additionally, composites of disulfide/polydicyclopentadiene (MoS₂/PDCPD) were prepared through *in situ* ROMP by a simulated RIM device (figure 5A). The production of glass fiber reinforced composites was also carried out in a simulated RIM mold (Figure 5E). Before polymerization started, glass fiber mesh was firstly arranged in the mold, where fabricated PDCPD composite were then produced. Consequently, the modified composites exhibited enhanced tensile, flexural and impact strengths. For the purpose of rheological measurement, the RIM apparatus was connected to a viscometer, a rheometer, a thermocouple, and a computer (Figure 5B).[126a,b,c,d,129]

Recent developments in this field have led to a renewed interest in new RIM mold materials. Hot injection molding molds made of traditional aluminum metal alloys showed good performance due to aluminum's anti-adhesive properties. Cold injection molding processes, where reactions were protected under nitrogen, required less energy consumption.[124] However heat was found to be essential to accelerate thermal crosslinking and to avoid the formation of defects or voids due to rapid polymerization. In addition, the use of low injection pressure is advocated to reduce long-term damage to the molds (Figure 5C, D and F). The resulting products had accurate geometric shapes if heating was used in the RIM process.[123] A 3-D printed polylactic acid (PLA) was used as a mold material and solvent was added to the mold in order to prevent external heat during processing. When the maximum reaction temperature reached 211 °C in a short period of time, PLA deformed in this extreme condition. Even though PLA was selected as a replaceable mold material at the beginning, the mold was then switched back to aluminum later in the study.[124]

1.4.0 PDCPD applications
1.4.1 PDCPD based composites

Recent studies have revealed that a wide range of materials can be used to make PDCPD based composites. Several approaches were used to accomplish the enhancement of mechanical properties. For example, triblock-copolymers poly(lactide)-*b*-poly(*p*-norbornenylethylstyrene-s-styrene) PLA-*b*-P(N-*s*-S) were mixed with dicyclopentadiene in the presence of G2 to afford novel composites.[95] The resulting materials had well-defined nanoporous structures and displayed remarkable mechanical strength, good thermal stability and permeability.[95] Another study was carried out in a similar copolymer system, which demonstrated that fabricated cylindrical monoliths of polylactide-*b*-poly(norbornenylethylstyrene-*s*-styrene) PLA-*b*-P(N-S) were blended with DCPD. The nanoporous composites exhibited comparable tensile strength to the PDCPD homopolymer.[102]

Emulsion ring opening metathesis polymerization can also be used with G2 for PDCPD/PNB synthesis. After being blended with well-dispersed carbon nanotubes (CNT), this copolymer exhibited exceptionally high compatibility in aqueous solutions. Electrically conductive PNB-*co*-PDCPD/CNT composite nanoparticles also showed excellent thermal stability and optical properties that could be potentially used in electronic devices.[101]

A considerable number of papers have been published on homopolymer and copolymer PDCPD based foams.[142a-f,144] These foams were made by a high internal phase emulsion (HIPE) technique. Polymers that are produced from HIPE are called polyHIPE, which are polymer particles in colloidal dispersions with high (normally more than 70%) internal phase volumes.[144g] These novel materials have designable crosslinked networks. Also, advanced modification of PDCPD based composites can be made such as designing microporous structural materials to afford good permeability and to tailor functionality.[142a-f] For example, a composite that consists of titania and PDCPD was prepared *via* HIPE by adding surface modified TiO_2 nanoparticles. It demonstrated potential photocatalytic activity, therefore, the composite was used as a supportive material for photocatalytic water purification.[110c,131] The polyHIPE foam swells in toluene and recovers to its original shape after being dried.[110b,c] Additionally, supercritical CO_2 was applied to dry out the foams to minimize the usage of organic solvent.[144e] Stabilized surfactants were used during the formation of the polyHIPE films and introduced ductility to the foam at the micron scale.[142b,e] Another experiment indicated that ZnO particles along with oleic acid together acted as a stabilizer. They were coated to the HIPE produced materials.[142f] To enhance the microstructures, a ruthenium based complex was used as an initator to crosslink the polyHIPE networks.[142b] In many cases, an antioxidant was added periodically to prolong stability. This also helped to improve certain properties like compressive strength.[142c] In addition, one study that was published in 2020 demonstrated that the polyHIPE templated PDCPD foams exhibited Young's modulus up to 2.1 GPa, electronic conductivity up to 2800 S/m and more than 97 % carbon content.[142a] The development of PDCPD based polyHIPE materials tends to applications of self-decontaminating filter media for use in deactivating chemical warfare agents.[144a]

Several studies have revealed that the PDCPD HIPE templates can also be used to stabilize *Candida* Antarctica lipase B in waste stream upcycling (Figure 6). The immobilized lipase demonstrated improved and repeatable activity up to at least 10 cycles. This template, a hierarchical material, can withstand industrial harsh conditions such as toxic solvents and extreme temperatures.[157c,d] In earlier works, DCPD was coupled with Grubbs catalyst, iron oxide nanoparticles, and lipase B to form oil microcapsules.[157a,b] The hierarchically self-assembled oil microcapsule had a rigid shell due to crosslinking of PDCPD. Hence, it tolerated acidic and basic conditions while immobilizing enzyme stability.[157a,b] Furthermore, copolymers of norbornene and dicyclopentadiene were applied as separation membranes in lithium-ion batteries.[166] In another application, brominated polyHIPE foams gained inductive-heating capability when mixed with iron surfactant nanoparticles. The resulting composite materials exhibited magnetic properties with cellular morphology and good mechanical properties.[144d] Additionally, superparamagnetic magnetite/polymer composites were made by mixing the ferrofluids.[144b,f]

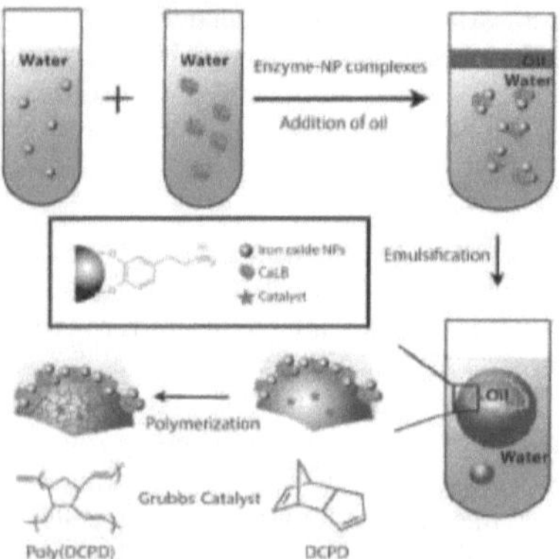

Figure 6. Schematic illustration of the crosslinked lipase microparticle (CLMP) assembly. Dicyclopentadiene was encapsulated with Grubbs catalyst and enzyme nanoparticles. Adapted from literature.[157]

Recently many studies have been focused on glass fiber blended PDCPD composites. These materials exhibited improved impact strength, extended fatigue life and enhanced compressive strength.[143a,b,c] Different types of fiber content and decelerator solutions (which were applied in order to slow down reaction and curing process) have effect on impact strength and ductile to brittle temperature (DBTT) of composites.[143e] Additionally, according to a hygrothermal aging test, PDCPD composites showed reduced water absorption relative to the corresponding homopolymers or to standard epoxy resin when immersing into water. That is because their intrinsic hydrophobicity and less degradation rate comparing to equivalent epoxy resins. After being immersed in water for a year, PDCPD still showed low absorption and no plasticization effect.[143d] Consequently, this result suggests the application of PDCPD and its composites towards service in humid environments.

Numerous studies have shown that PDCPD/silica nanocomposites exhibited excellent electrical properties like low dielectric constant, low dielectric loss and good corona resistance.[145a,b,c] These nanocomposites also had outstanding mechanical properties. For example, the nanocomposites showed excellent break down strength, flexural strength, flexural modulus and coefficient of linear expansion, and these parameters were at the same level as commercial electric insulating material like bisphenol A type epoxy resins.[145b] Hence, the potential applications of silica impregnated PDCPD nanocomposites include high frequency and high voltage materials.[145a,b] Furthermore, surface modified PDCPD/BaTiO$_3$ nanocomposites exhibited enhanced dielectrical properties and break down strength. As a consequence, these novel materials might be applied in energy storage devices.[145c]

Recently, a significant enhancement in the toughness of composites was achieved by establishing a covalent bond between the crosslinked PDCPD network and vinyl-functionalized SiO$_2$.[146a] Perhaps thanks to the presence of microvoids and microcracks, which can absorb external energy when force is applied, several mechanical properties were toughened. These microstructures that were formed between the silica and carbon-based networks contributed to energy absorption and dissipation when external forces were applied.[146a] Another approach to enhance such properties was to add phenyl functionalized silica (P-SiO$_2$) particles. The functional group on the SiO$_2$ particles added as a filling agent had an effect on the enhancement of the nanocomposite's mechanical properties. The rigid phenyl group efficiently enhanced the yield strength and notched Izod impact strength of the material.[146b] In addition, such factors like filler loading, filler/polymer compatibility, and filler functionality affected the interfacial bonding of materials.

Other similar studies have successfully demonstrated that PDCPD/silica nanocomposites could be prepared by *in situ* polymerization. Nanocomposites that were prepared within RIM molds demonstrated dramatically increased mechanical and tribological properties. In fact, nano-scaled silica particles are advantageous to improve wear resistance, hardness and bending strength of the PDCPD matrix.[146c,d,e,f] Moreover, other strategies were applied in order to enhance the mechanical behavior and swelling ratio of the composites. These strategies include combining multi-walled carbon nanotubes (MWNTs), functionalizing the MWNTs with norbornene or benzoic acid, and adding rigid additives like phenolphthalein-containing bismaleimide (PPBMI).[147a,b,c,d] Typical universal mechanical tests such as tensile strength test, impact strength test, dynamic mechanical analysis, and notched Izod impact tests were performed. All results were proven to be positive. In some extreme cases, the tensile toughness was dramatically enhanced to 900 % of the parent PDCPD.[147b,c] Another method to improve the mechanical and tribological properties was to mix PDCPD with surface modified nanoparticles. For instance, nanoparticles like MoS_2 were hybridized with dialkyldithiophosphate (PyDDP) then were mixed with PDCPD to form nanocomposites.[163]

Several studies have revealed that graphite or vapor-grown carbon fibers (VGCF) that were combined with composites exhibited reinforced properties due to the well-established networks between crosslinked PDCPD backbones and mildly oxidized graphene oxides (MOGO).[151a,b] Additionally, nanocomposites containing inexpensive montmorillonite (MMt), which is a layered crystalline mineral extracted from clay, were popular additives. These nanocomposites were prepared by WCl_6/Et_2AlCl through ROMP or by simply mixing pre-polymerized PDCPD with poly(ethylene glycol) PEG.[155] The results from X-ray diffraction (XRD), TEM and small angle neutron scattering (SANS) indicated that the delamination led to nanodispersion into PDCPD matrices. As a consequence, the thermal stability was increased dramatically as MMt loading increased.[155a,b,c] Surfaces of these nanocomposites can be eroded by oxygen plasma to obtain particles with oriented layers.[155d] Many studies have indicated that strong interactions between ultra-high molecular weight polyethylene fibers and PDCPD matrices resulted in the adhesive interfaces.[158a,b]

1.4.2 PDCPD based aerogels

Aerogels, porous materials that have low density and high surface area, exhibit optical, dielectric and acoustic insulating properties. Aerogels are formed by replacing the solvent that is trapped within a PDCPD gel with air at atmospheric pressure. By contrast, gels that are produced through an ambient pressure drying process are xerogels. PDCPD based gels were discovered in 2007.[133] This novel aerogel material has attracted much more attention over the past 10 years. As a new insulating material, it has light weight, nanoporous structures, thermal stability, and excellent mechanical properties while it costs less than traditional aerogels. Typical procedures for making PDCPD based aerogels are begun with dissolving solid dicyclopentadiene at room temperature under atmospheric pressure in a less polar solvent like toluene, 1-butanol or 2-propanol with stirring. Then ruthenium catalyst solution is added to this colorless transparent precursor solution, resulting in the formation of a PDCPD gel. Next, wet gels were aged in acetone for 1–2 weeks, followed by washing with acetone to remove unreacted monomers. This method is known as sol-gel processing. To form aerogels, wet gels are dried with supercritical CO_2 to extract residual solvent, then dried in an oven at 50 °C. Characterization of the product aerogels included kinetics of gel formation, linear shrinkage and thermal conductivity. These new gels are homogenous with high nanoporosity ratio which results in inherent hydrophobicity.[133]

Many studies have shown modification methods for aerogel and xerogel systems. For instance, to increase oxidative stability, aerogels were hydrogenated. The hydrogenation resulted in increased glass transition temperatures.[139] One study on improving gel formation was to dry wet gels with liquid CO_2 rather than CO_2 gas. This methodology was applied to fabricate PDCPD-*ran*-PNB copolymer aerogel layers in hollow spheres.[134]

Several studies have been carried out on the improvement of the catalyst system to optimize gel properties. As an illustration, PDCPD aerogels that were catalyzed by G1 displayed well-shaped monoliths but the gels produced by G2 were deformed. However, G2 produced aerogels were more squeezable and less rigid.[135a] SEM/TEM data indicated that when small PMMA particles were added to the system, G2 produced aerogels obtained higher mass-fractal aggregated particles which were partially interpenetrated into each other.[135a,136a,b] In addition, WCl_6 and a ditungsten cluster were used to control the stereoselectivity of PDCPD polymerization. The *cis/trans* ratio is related to the swelling ratio of gels. In terms of stereo-geometry, the gels that were made by tungsten complexes had higher *cis/trans* ratio than ruthenium catalyzed gels. Thus, the tungsten-based complex catalyzed gels gained higher swelling ability than the ruthenium catalyzed gels.[135b] Likewise, PDCPD xerogels were also produced by a ditungsten cluster containing mostly *cis* configuration. These xerogels exhibited extreme swelling behavior in a diversity of organic solvents (aliphatic and aromatic solvents) and were used as templates to separate organic solvents and water.[135c]

When partially crosslinked PDCPD was grafted with linear PMMA, the resulting composite material was able to resist deformation and swelling due to the precise-designed nano-topological networks between PDCPD and PMMA. Consequently, less deformed and well-controlled nanostructures resulted in enhanced physico-mechanical properties, so that the copolymer gels can be essentially used as shape memory and chemical delivery devices.[136a,b] Additionally, copolymers consisting of norbornene derivatives were reacted with PDCPD as crosslinkers to form rigid (low swelling ratio) but extra light aerogels.[137] One study has clarified that in order to be applied as coating materials, the rheological properties of copolymers have to be clearly studied.[138a] In addition, an approach to controlling gelation and viscosity in the gel system was demonstrated through reduction of crosslinking. This was achieved by copolymerizing a linear polymer into the system. For example, doped (halogenated) PDCPD aerogels with high Z tracer elements were used in the fabrication of ignition targets for inertial confinement fusion (ICF).[138b] Similarly, another study demonstrated that the (bis)iodo-doped copolymer PDCPD-*ran*-PNB-I_2 adhered evenly in the hollow spheres of millimeter scale inner diameter of diamond inner shells.[162]

One study has demonstrated that the non-covalent metal coordination can be used as the basis for a reversible network to crosslink soluble linear PDCPD (Figure 7). This enabled the easy-accomplished synthesis of chemo-reversible and highly tunable metallogels based on polydicyclopentadiene. Polymer reclamation and metal coordination were repeated without degradation. Also, by altering the type of metal ion, metallogels exhibited tunable dynamic mechanical properties and excellent chemo-reversibility.[140] Therefore, a precise method for making uniform aerogel coatings was well established.

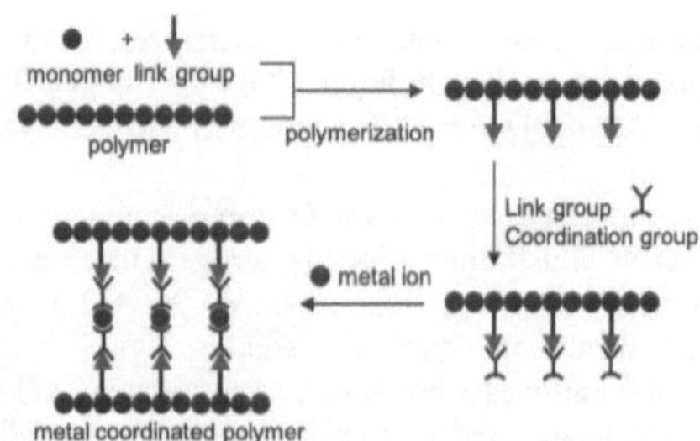

Figure 7. Schematic demonstration of metal-coordinated non-covalent crosslinking networks through linear chains. Adapted from literature.[140]

1.4.3 PDCPD based self-healing materials

The idea of developing self-healable polymers was proposed in 1980s, demonstrating the property of such a polymer that could self-repair spontaneously after it receives external damage.[178] In 2001, the Moore and White groups collaborated on the development of a novel autonomic healing composite system.[169] This site-specific self-repairing composite contained an epoxy matrix that was mixed with Grubbs catalysts and DCPD microcapsules (Figure 8). Theoretically, once external forces are applied to the system that cause cracking, microcapsules would break and release the DCPD monomers. Monomers react with the catalysts that are outside of the microcapsules to form crosslinked PDCPD networks. As a result, healed materials recovered 75 % of the original fracture load.[169] Although this work demonstrated a solution to solve existing problems like microcracking and damage of polymeric materials, it did not fully explain the healing kinetics and method of improving catalyst stability. Afterwards, PDCPD became a popular target in the development of self-healing materials.

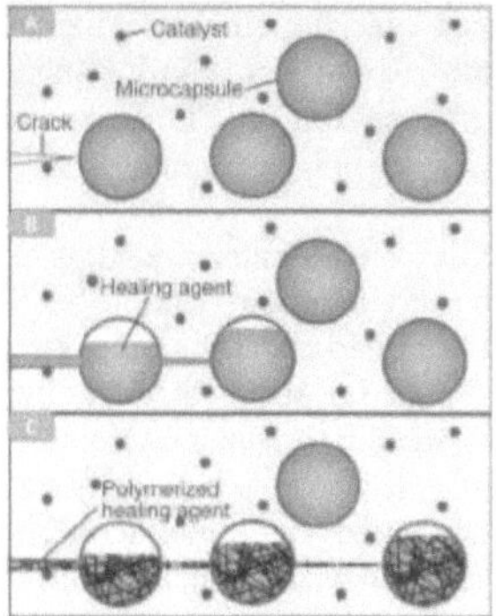

Figure 8. The demonstration of autonomic healing concept. A microencapsulated healing agent is embedded in a structural composite matrix containing catalyst that is capable of polymerizing the healing agent. A) Cracks form in the matrix wherever damage occurs. B) The crack ruptures the microcapsules, releasing the healing agent into the crack plane through capillary action. C) The healing agent contacts the catalyst, triggering polymerization that bonds the crack faces close. Adapted from literature.[169,176]

Several subsequent papers focused on modification of the self-healing system. Rheokinetic studies on Grubbs catalysts based self-healing materials have indicated that the morphology of catalysts influenced the dissolution kinetics as smaller crystal sizes led to faster dissolution rate and higher healing efficiency.[170a,c] G1, G2, and HG2 are the most commonly used catalysts for the homo and copolymerization of PDCPD. In a comparative study, G2 produced polymers showed the best thermal stability and fastest initiation rate. In the meantime, all three catalysts exhibited similar chemical stability and functional group tolerance.[170b,176] In addition, the G2-initiated polymer had higher modulus value and longer gelation time.[170b]

Many studies have demonstrated the use of wax to protect DCPD microcapsules, while switching the matrix to epoxy vinyl ester. Consequently, this modified material gained optimized mechanical properties and healing efficiency. In order to move towards commercializing PDCPD based self-healing materials, a low-cost and environmental stable tungsten complex was selected as a replacement for the expensive ruthenium-based catalysts. One recent study has demonstrated the use of polystyrene (PS) fibers that were processed *via* electrospinning as a novel matrix. At least 90 % energy was recovered after several repeating cycles, according to the impact strength test. The is because fabricated PS leads to better catalyst dispersion.[175] Another study compared the effect of three different forms of WCl_6 including recrystallized, wax protected and as received. These methods resulted in different microcapsule size distributions. This stable tungsten-based precursor that was combined with a coactivator phenylacetylene and coupling agents showed positive results in the polymerization of self-activated *exo*-dicyclopentadiene composites. Furthermore, statistical data indicated that the self-healed polymers exhibited 75 % toughness of neat resin and good healing efficiency.[171]

Many efforts have been made on the optimization of the healing agents among the norbornene based homo-copolymers. For instance, a library was built up that had 13 ROMP-active monomers including DCPD. This small library study enabled the prediction of Grubbs catalysts solubility in liquid monomers.[172a]

Moreover, several studies have demonstrated strategies to increase healing efficiency. Rheokinetic results on homopolymerization implied that 5-ethyledene-2-norbornene (ENB) and *exo*-DCPD exhibited faster initiation rate than the *endo*-DCPD.[172b] In addition, *exo*-DCPD containing material self-healed 20 times faster than the *endo*-DCPD but resulted in 20 % lower healing efficiency. Dimethylnorbornene ester (DNE) that was copolymerized with DCPD formed a new self-healing system. This novel self-healable network was based on noncovalent interactions, including hydrogen bonding. As a consequence, healing efficiency and mechanical properties were significantly improved within these blends' composites.[173a]

The newest designed DCPD self-healing system can be used as an adhesive material. When 5-norbornene-2-methanol (NBM) was added to DCPD/ENB as an adhesion promotor with Grubbs catalysts, this hydrogen bonded self-healable system showed strong adhesion characteristics.[173b] Notably, this copolymer system had 72 % healing efficiency. Another study has demonstrated a reversibly polymerizable homo-comonomer system based on cyclopentene derivatives.[122d]

1.4.4 PDCPD based copolymers and polymer blends

Recent developments have indicated methodology for enhancing physicomechanical properties of PDCPD based copolymer systems . For example, a norbornene derivative 8-methyl-8-methoxycarbonyltetracyclo[4.4.0.1^{2,5}.1^{7,10}]dodec-3-ene (MMT) and DCPD were combined together through ROMP to form copolymers. These copolymers had similar glass transition temperatures to PDCPD but showed enhanced properties such as anti-water absorption and toughness. Some of these enhanced properties were attributed to the polar group in the MTT co-monomer.[96] In addition, DCPD was exploited as a crosslinking agent during a copolymerization with a norbornenyl-functionalized castor oil alcohol to increase glass transition temperature and improve storage modulus.[97] Specifically, copolymer blends of DCPD and Dilulin™ ([3-[(*Z*)-octadec-9-enoyl]oxy-2-[8-[3-[(2*Z*,5*Z*)-octa-2,5-dienyl]-2-bicyclo[2.2.1]hept-5-enyl]octanoyloxy]propyl](*Z*)-octadec-9-enoate) which is a drying oil-based reactive diluent, gained a tunable glass transition temperature and crosslinking density. Dilulin (which is synthesized using a Diels-Alder reaction) has long carbon chains. These flexible linear chains act as plasticizers during PDCPD processing and therefore introduce flexibility into the system.[98a,b] In addition, copolymer thin films of tetraalkylammonium-functionalized norbornene and dicyclopentadiene exhibited high hydroxide ion conductivity and good mechanical properties. These thin films that were processed through a facile ROMP route also have exceptional methanol tolerance.[100] Some other blends consisting of norbornene derivatives like 1,4-dicyclopentadienylmethyl benzene (DMB) or 2,2'-dicyclopentadienyl ether (DCPE) had similar crosslinking behaviors as PDCPD but enhanced thermal stability and toughness.[164] A rigid aromatic compound like phenolphthalein-containing bismaleimide introduces extra rigidity into the

PDCPD. Hence, this PDCPD blend showed improved thermal stability and bending strength.[165]

Many studies on thermoset/thermoset blends of PDCPD and epoxy showed appealing results. These blends have tunable nanostructures and exhibit enhanced mechanical properties. Phase separation of blends, which took place due to different crosslinking rate of PDCPD and epoxy, was controlled by adding an amphiphilic block copolymer poly(1,4-butadiene-*b*-ethylene oxide) without sacrificing the optimized strengths.[109a,b,c,d] According to the SEM images, ultra-small-angle and small angel X-ray scattering analysis, the morphology of the blends exhibited fractal structures in both PDCPD rich and epoxy rich phases. The fractal structure dissipates a large amount of external energy before fracture. Owing to the strong bond strength between blends, the materials can be applied to adhere carbon fiber composites to steels.[141a,b,c]

Several studies have revealed that a comonomer can act as a crosslinker that takes part in the polymerization. It is possible to control the thermal properties and overall mechanical behaviors of PDCPD through this strategy. The resulting copolymers had a wide range of rigidities and stiffness in severe service conditions.[150a,b] Furthermore, DCPD was used as a crosslinker in copolymerization with benzenesulfonyl chloride-functionalized norbornene to make a proton exchange membrane with high ion exchange capacity (IEC) and high proton conductivity.[167]

Hydrogenated PDCPD was blended with long chain branched polyethylene and conventional polypropylene in order to form tensile biaxially oriented polypropylene films. Notably, these stretchable films have high Young's modulus (4.9 GPa).[161] In addition, crosslinker monoterpenes (limonene oxide, d-limonene and -pinene) were added to the PDCPD homopolymers. The addition of crosslinker enabled secondary metathesis and chain transfer events to occur, in order to solidify the resulting copolymers. Consequently, this method led to the formation of a hyperbranched structure that has dendritic units.[179]

The most recent studies in this field emphasized a renewed interest in developing recyclable and degradable thermoset materials. For instance, the Johnson group developed re/upcyclable and degradable thermoset copolymers that were polymerized from dicyclopentadiene and silyl ether monomer (Figure 9). The cleavable network was well established with crosslink sites from cyclopentene double bonds and reversible covalent bonds like Si–O bonds. This low-cost comonomer strategy enables the recyclability and degradability of high performance thermoset materials.[187]

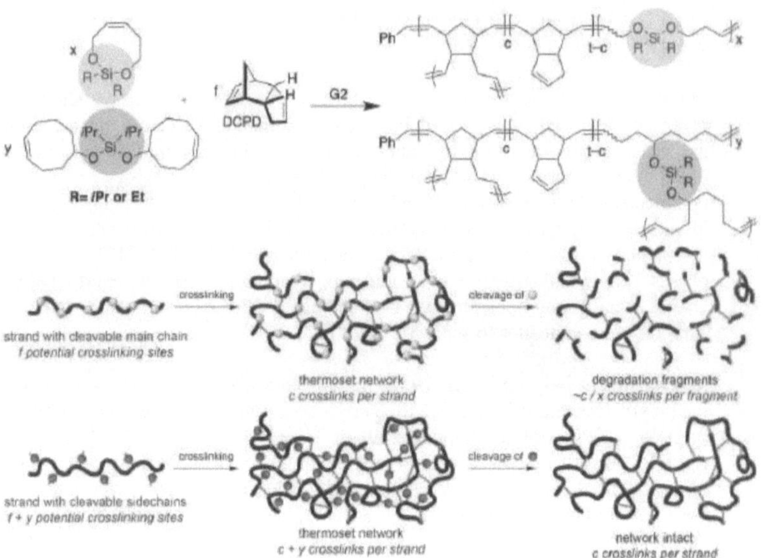

Figure 9. Copolymerization and crosslinking of PDCPD with silyl ether groups. Adapted from literature.[187]

1.4.5 PDCPD in industry

Having been obtained as a by-product from naphtha crackers, *endo*-DCPD has attracted much interest in industry as it has been polymerized into PDCPD.[125] Indeed, PDCPD has used in the automotive industry for several decades. For example, for the purpose of replacing aluminum alloy while maintaining the same level of strength and vibration damping, PDCPD was designed to be a new generation of truck body panels as well as car power steering housings. In addition, PDCPD that was mixed with isocyanate foam has potential application for power steering housings.[123,124]

PDCPD has also been applied in marine devices, recreational vehicles, and heavy vehicles. It has been used as a tonneau cover for pick-up trucks, snow mobile hoods, sporting goods, tractor fenders, and auto bumpers.[125] Several studies have shown that PDCPD fibers possessing reduced dimensions into microscale as a result of electrospinning exhibited slightly higher Young's modulus value than the corresponding bulk materials. Additionally, millimeter scaled hollow fibers were used as coolant channels for cylindrical type lithium-ion cells.[149]

One study of PDCPD durability in seawater at high temperature and pressure was carried out. The result indicated that PDCPD is a suitable coating material for offshore steel pipes.[143d,144c,150b] Comparing to conventional coatings like polypropylene and polyurethane, PDCPD has good thermal insulation properties, higher operation temperature (maximum temperature range is 180–204 °C), lower water absorption (than epoxy), good oxidative stability and excellent adhesion. Because of its low viscosity, this tough resin is able to penetrate into rocks and adhere to the rock when once it is crosslinked inside the porous structure. PDCPD can therefore be used for sealants and plugs under water.[150a] In addition, crosslinked PDCPD films can separate organic molecules in the range of 100–600 g/mol. The permeated nanofiltration membrane filters both polar and apolar molecules depending on the cross-sectional area.[159ab] More specifically, *trans*-fatty acids and saturated fatty acids, when added together with triisobutylamine, permeate faster than *cis*-fatty acids through the PDCPD based membrane. That is because triisobutylamines react with fatty acids to form noncovalent bonds and the reacted *cis*-fatty acids obtained larger critical areas than the saturated and *trans*-fatty acids. Hence, fatty acid is separated stereoselectively. This nanofiltration can be applied in industry at large scale.[159,182]

1.5.0 PDCPD functionalization and modification
1.5.1 Surface functionalization and modification of PDCPD

There are a few reports that describe methods for functionalization of the PDCPD surface through reaction of the double bonds within the polymer. For example, one study demonstrated a method for grafting thiol-linked methylmethacrylate groups onto the olefinic backbone through a thiol-ene reaction. Polymerization through these groups resulted in grafted PMMA layers (Scheme 9). Notably, the PDI of the grafted PMMA polymers was controlled *via* atom transfer radical polymerization (ATRP).[180] In another study, the PDCPD surface was brominated then incubated in 4-(trifuoromethyl)benzylamine in *N,N*-dimethylformamide (DMF) for the purpose of monolayer formation. Both brominated or amine substituted PDCPD surfaces can be applied in biological devices or microfluidics.[181] Additionally, PDCPD surfaces that were reacted with *m*-chloroperoxybenzoic acid allowed the incorporation of epoxides at nanometer scales.[148] The surfaces could also be modified by norbornene-functionalized polyethylene glycol telechelic oligomers to afford nanoporous structures.[160]

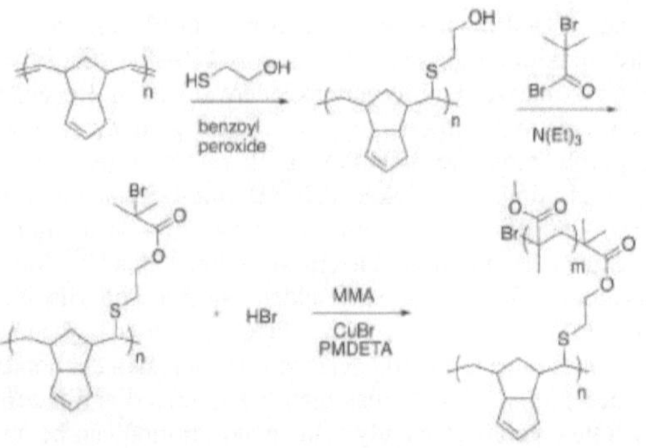

Scheme 9. Previous works on functionalization of PDCPD surface. Adapted from literature.[180]

1.5.2 Early approaches to functionalized monomers

Dicyclopentadiene has an unpleasant odor, which is detectable by humans at a concentration of 5 ppm. The presence of residual monomer within PDCPD therefore limits possible applications of the polymer.[184a,d] One possible approach to solving the odor problem is to chemically modify the structure of the DCPD monomer. One way, for example, is to biologically oxidize and partially degrade the polymer by microorganisms. This reduces the average molecular weight of the polymer and removes the objectionable odor, presumably through co-degradation or extraction of residual monomer.[156] However, the biodegraded PDCPD loses the excellent mechanical properties that make PDCPD valuable in the first place. Notably, modified PDCPD that was produced by PolyHIPE technology and treated with ionizing radiation maintains reasonable physicomechanical properties.[132]

An alternative strategy would be to alter the structure of the DCPD itself, in such a way that the monomer loses its unpleasant smell, but retains the ability to undergo polymerization. For example, the Lemcoff[184b] and Xu[183] groups carried out an allylic oxidation to regioselectively afford an alcohol-functionalized monomer that could be reacted with a variety of electrophiles to produced different substrates that contained new functionality at the allylic position (Scheme 10). The Xu group used the G1 catalyst to produce a linear acetoxy functionalized PDCPD. The manuscript describing this work reported a glass transition temperature in the range of 136 °C–159 °C, but presumably such a high T_g indicates that some degree of crosslinking occurred during the thermal analysis. Xu's polymer decomposed above 221 °C, which indicates substantially reduced thermal stability (which is typically stable up to almost 400 °C) relative to the parent PDCPD material. Presumably, the reduction in thermal stability originates from the presence of the labile allylic ester group. Similarly, the Lemcoff lab tested various functional groups at this position including the original alcohol, a methoxy derivative, and several esters. However, none of these functionalized polymers had comparable thermal stability to the original PDCPD.[106c,183,184a]

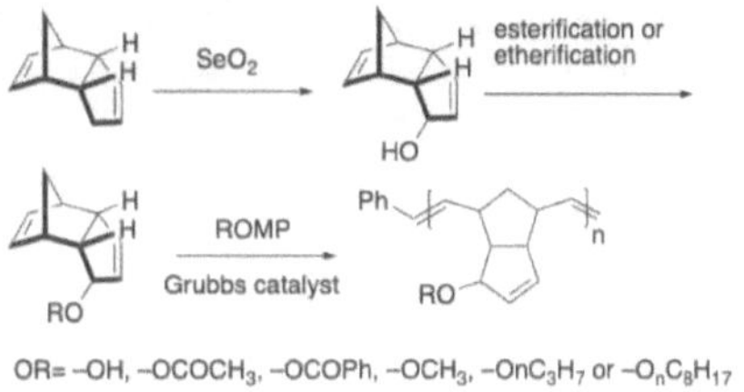

Scheme 10. Lemcoff and Xu's studies on functionalization of DCPD and PDCPD. Adapted from literature.[183,184a,b]

1.5.3 Ester functionalized polydicyclopentadiene (*f*PDCPD) incorporating a methacrylate motif

In order to incorporate new functionality within polydicyclopentadiene without reducing the thermal stability of the polymer, a previous PhD student within the Wulff lab—Dr. Jun Chen—designed an improved monomer that links an ester group through a strong sp^2–sp^2 bond onto the non-strained alkene within dicyclopentadiene.[191]

Synthetic access to this new monomer stemmed from the Wulff lab's prior research into the chemistry of so-called Thiele's esters, which are doubly carboxylated dicyclopentadiene derivatives produced through regioselective Diels-Alder reactions.[194,199a,b,239] The presence of the ester group confers several useful properties to both the monomer and the resulting polymer. First, of course, it provides a synthetic handle that can be harnessed to alter the surface energy of PDCPD or potentially used to attach new functionality. Additionally, the presence of this electron-withdrawing group on the pendent cyclopentene ring that remains after the ROMP process has taken place diminishes the likelihood of secondary metathesis events. This allowed for a facile isolation of the linear polymer, which could then be thermally crosslinked in a separate operation. Finally, the ester-functionalized monomer had a fruity odor which was much more pleasant than the foul smell of the unmodified dicyclopentadiene.

The synthesis of ester-containing dicyclopentadiene monomer **4** begins by converting sodium cyclopentadienylide **1** to a stable metal salt intermediate **2** through the addition of five equivalents of dimethyl carbonate. Protonation to generate the neutral form of intermediate **2**, followed by Diels-Alder reaction with dicyclopentadine afforded a mixture of the desired functionalized monomer (**4**), along with dicyclopentadiene **6**, Thiele's ester **5**, and non-polymerizable regioisomer **3** (Scheme 11). The unwanted dicyclopentadiene and Thiele's ester could be removed from the reaction mixture, at which point the mixture of **3** and **4** were exposed to second-generation Grubbs catalyst to produce the desired linear polymer (**7**). Unreacted **3** was recovered from the supernatant after precipitation of the polymer. Thermal curing produced a crosslinked form of the polymer.

This new form of functionalized PDCPD (*f*PDCPD) displayed the highest T_g (172±3 °C) among all functionalized and unaged PDCPDs described to date. Also, owing to the presence of the ester group, the hydrophobicity of the *f*PDCPD surface could be tuned through partial hydrolysis. The surface contact angle of *f*PDCPD films dramatically changed from 87.2±0.9° to 29.0±1.1° following saponification.[191] These data suggested the potential utility of *f*PDCPD in bio-material applications and polymeric microfluidic devices.

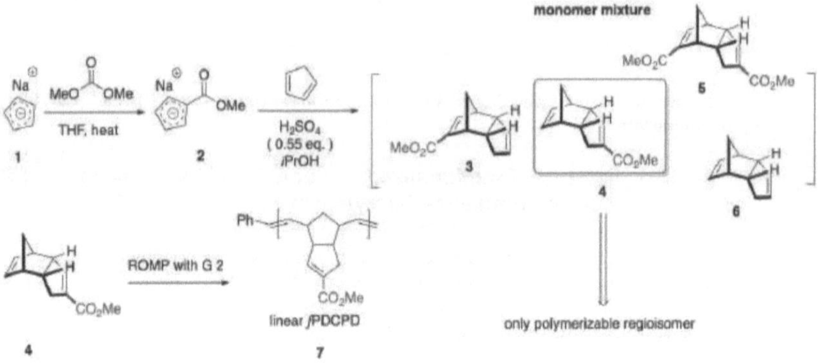

Scheme 11.Synthetic route of methyl ester functionalized dicyclopentadiene (*f*DCPD).[191]

1.6.0 book objectives and summary of experimental chapters

Functionalized polydicyclopentadiene (*f*PDCPD) readily undergoes crosslinking when exposed to air, and the rate of crosslinking increases at high temperature. To understand the mechanism of crosslinking and establish the structure of the chemical crosslink formed with *f*PDCPD, we prepared both deuterated and non-deuterated linear *f*PDCPD, and proceeded to crosslink the samples *via* thermal annealing at various temperatures (normally 60–180 °C). FTIR, Raman, and Solid-state NMR spectroscopies were used as characterization techniques. From these data, we conclusively established that the mechanism of thermal crosslinking within *f*PDCPD is predominantly head to tail olefin addition from one methacrylate group to another. Oxidative crosslinking through an autoxidation mechanism was identified as a secondary process.[192] (chapter 2)

We next improved the synthesis of the *f*DCPD monomer mixture on a half-kilo scale with minimal use of solvents. We also developed an in-lab reaction injection molding process and successfully manufactured thermally crosslinked samples for different purposes.[193] Dynamic mechanical analysis was used to investigate the mechanical properties and glass transition behaviors of the material. Significantly, our results reveal for the first time that our ester-functionalized PDCPD has equivalent mechanical strength to that of the parent polymer materials.[193] (chapter 3)

We expanded on the utility of *f*PDCPD by harnessing the embedded functional group to manipulate the surface energy of the material, and to attach a range of biologically relevant functional groups. A fluorescent dye was firstly used to characterize the surface coverage, and an RGD peptide was incorporated to promote adhesion of mammalian cells. Following these experiments, the antibiotic drug chloramphenicol was attached through the ester group, to create an environmentally responsive polymer that was able to moderate *E. coli* growth. These results constitute the first report of a functionalized PDCPD that has surface bioactivity.[196] (Chapter 4)

Finally, statistical polymers were synthesized with various molar fractions of functionalized and non-functionalized monomers. The chemical structures were investigated using solution and solid-state NMR, along with linear vibrational spectroscopy (FTIR and Raman). Contact angle measurements revealed tunable hydrophobicity for the polymer surfaces. More importantly, Vickers hardness data and DMA results indicate that the copolymers have equivalent mechanical properties to the original PDCPD.[195] (Chapter 5)

Chapter Two: Structure of the Thermally Induced Crosslink in *C*-Linked *f*PDCPD

The material in this chapter was adapted from: "Structure of the Thermally Induced Crosslink in *C*-Linked Methyl Ester Functionalized Polydicyclopentadiene (*f*PDCPD)" T. J. Cuthbert, **T. Li**, A. W. H. Speed, and J. E. Wulff, *Macromolecules.* **2018**, *51*, 2038–2041.[191]

The synthesis of compounds **4** and **7**, together with isotope labeling of compounds **15**, **16, 17**, and **18**, and their characterization by NMR and FTIR were accomplished by Dr. Tyler Cuthbert. The reduction to afford compound **19** was accomplished with the help of Prof. Alexander Speed at Dalhousie University. Solid-state NMR measurements were carried out by Dr. Mathew Willans at Western University. The FTIR and Raman spectroscopy measurement, sample preparation and data analysis were accomplished by **Tong Li.**

2.1.0 Overview

Understanding the chemical nature of the materials that we produce is necessary to achieve a fundamental understanding of the physical behaviour of a given material and is useful in predicting how we might rationally manipulate its properties. For novel thermosetting materials, determining structure-property relationships is particularly challenging, because of an inability to conduct absolute characterization using conventional methods of the type employed for soluble (macro)molecules with uniform chemical structures. Unfortunately, this often leads to assumptions about materials produced that could ultimately affect applications, further functionalization, and modification.

Polydicyclopentadiene (PDCPD) is a thermoset polymer that is generated by ring-opening metathesis polymerization (ROMP) from dicyclopentadiene (DCPD **6**; Scheme 12A). Thanks to its low cost, high strength, and corrosion resistance,[11,105] PDCPD has been broadly employed in the automotive industry, where it is used to make body panels, bumpers, and other components for trucks, buses, tractors, and heavy-duty construction equipment. More recently, the ability to generate porous PDCPD foams or to encapsulate microencapsulated PDCPD within other materials has suggested new applications in tissue engineering, gas storage and self-healing materials.[133,142b,d]

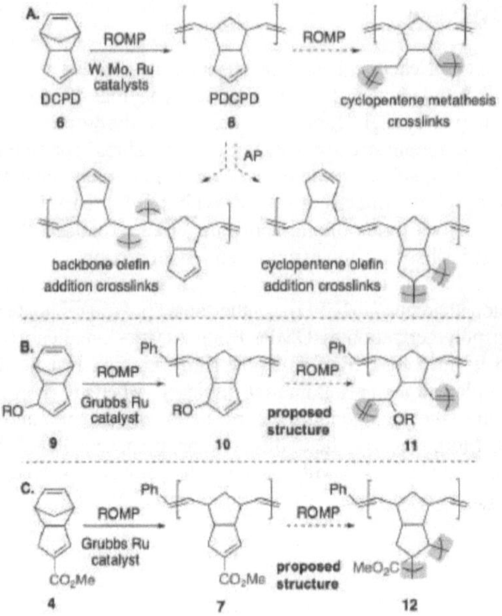

Scheme 12. Reported PDCPD crosslinks arising from different mechanisms: A) polymerization of unmodified DCPD, and potential crosslinks arising from subsequent ring opening metathesis polymerization (ROMP) or thermal addition polymerization (AP); B) work by Xu and Lemcoff on *O*-linked allylic ether and ester polymers; C) previous work from our group on a *C*-linked ester polymer. The current study aims to conclusively establish the structure of the crosslink in the ester-functionalized material.

2.1.1 Crosslinking of PDCPD

Polymerization of DCPD on industrial scale most commonly involves the use of a WCl_6/Et_2AlCl catalyst system, and is known to produce a tough, extensively crosslinked plastic.[11,105a] Many other metathesis catalysts have also been employed in industrial or academic settings, including molybdenum, ruthenium, tungsten, titanium, tantalum, and organic catalysts.[22,40,52,104,105b,142d,202] While a few examples of linear polydicyclopentadine are known,[52,93,105a,140,202] the vast majority of PDCPD material appears to be densely crosslinked, and indeed these crosslinks are understood to be responsible for most of the useful material properties discussed above. However, the chemical structure of the crosslinks has been the subject of some debate.

The earlier literature on PDCPD suggested that crosslinks arise principally through the metathesis of both olefins within the DCPD monomer.[104] This metathesis-crosslinked structure (highlighted in purple in Scheme 12A) is frequently invoked even in recent publications relating to PDCPD, [108a,111b,e,139,146a,147b,149a,164,181,] but this may not always be accurate. Important studies from Wagener and coworkers in the late 1990s indicated that for the classic WCl_6/Et_2AlCl system–as well as for a representative Schrock alkylidene– metathesis crosslinks of this type could be unambiguously ruled out.[105a,b] Instead, Wagener's studies pointed to a thermally-initiated olefin addition reaction as the principal mode of crosslinking. Because the linear polymer intermediate **8** contains both backbone olefins and internal cyclopentene olefins, it is likely that both motifs participate in these addition reactions. No attempt was made to distinguish between crosslinks arising from backbone additions (highlighted in green in Scheme 12A) and cyclopentene additions (highlighted in blue).

Wagener's analysis relied upon the fact that while the strained norbornene-type olefin in **6** is readily polymerized by the two catalyst systems discussed in the preceding paragraph, the less-strained cyclopentene olefin does not undergo metathesis under the conditions used for the polymerization. The situation becomes more ambiguous with ruthenium catalysts, however, since cyclopentene itself has been shown to be a competent monomer for polymerization by Grubbs-type ruthenium catalysts.[58,59,203] One therefore cannot a priori rule out the possibility of metathesis-type crosslinks for any PDCPD polymer produced with a ruthenium catalyst.

Indeed, working under the assumption that metathesis crosslinks dominate the structure of Ru-polymerized dicyclopentadiene, Xu and co-workers added an acetate group at the allylic position of the DCPD monomer (compound **9** in Scheme 12B; R=Ac) with the intention of disfavouring the crosslinking process.[183] The resulting polymer (made using G1 at 0 °C) was described as being non-crosslinked and yet the relatively high T_g values (136–159 °C) suggest that at least some degree of crosslinking likely took place. For comparison, typical T_g values are around 53 °C for linear PDCPD[202] and 155 °C for crosslinked PDCPD.[106d] Although not considered in Xu's report, it is possible that crosslinking occurred thermally, under the conditions of the T_g measurement, as we have described in our own work.[194] In contrast, Lemcoff and co-workers reported the polymerization of a very similar family of allylically functionalized monomers (compound **9** in Scheme 12B; R=various substituents), but concluded that these did undergo crosslinking (using G2 at 70 °C), despite the fact that most products had much lower T_g values than those reported by Xu.[184a] Based upon the observation of new signals in the IR spectra (assigned as acyclic alkene C–H stretches), Lemcoff's report invoked structure **11** in describing the connectivity of the final materials.

The allylically functionalized polydicyclopentadiene materials described by Xu and Lemcoff allowed some degree of tuning of both the T_g and surface energy relative to unmodified PDCPD but suffered from poor thermal stabilities—likely because the allylic ester or ether groups can decompose to carbocations at elevated temperatures. Seeking to make more robust functional materials (preferably with higher glass transition temperatures), our group therefore designed an alternative monomer (**4**, Scheme 12C) in which the dependent functional group was attached through a less labile sp^2–sp^2 bond.[193,204] We intended the resulting linear polymer **7** to favour crosslinking through a thermal process, wherein the embedded methyl methacrylate motif would facilitate self-initiating radical polymerization upon heating. While the very high T_g for our material (172±3 °C) was consistent with our proposed crosslinked structure **14** (the increased steric hindrance that comes with installation of a quaternary centre in the proposed structure was designed to increase the glass transition temperature relative to unmodified PDCPD), we could not rule out other possible structures.

The difficulties in understanding PDCPD crosslinks at a chemical level—despite the fact that the parent material has been used industrially for over two decades—underscore the challenges associated with characterizing insoluble thermosets. Nevertheless, there are useful spectroscopic methods that can be brought to bear in interrogating the structures of insoluble, nonhomogenous polymers. Techniques such as solid-state NMR, FTIR and Raman spectroscopy are all applicable to these types of materials. $^{13}C\{^1H\}$ solid-state NMR can provide clues as to the introduction (or loss) of functionality in non-soluble materials, including PDCPD.[205] Monitoring polymerizations by Raman and FTIR have also been popular among coating, ink, and 3D printing applications where conversion of functional groups defines the materials' performance and characteristics,[206-208] and has included PDCPD characterization.[108a,109b,135b,139,149a] These techniques are informative but often lack detail and resolution for definitive proof of the crosslinked structure when used individually. Together, however, they can contribute to a more thorough analysis.

2.2.0 Investigation of crosslinking of functionalized polydicyclopentadiene

In the current study, we sought to unequivocally determine the crosslinked structure of our *C*-linked methyl ester-functionalized polydicyclopentadiene (*f*PDCPD) by determining the functional groups involved in crosslinking and the various reaction pathways that could result within crosslinked material.

To this end we considered five possible crosslinking scenarios:

i.   reactivity through both the backbone alkene and pendent cyclopentene by olefin addition;
ii.   olefin addition through the cyclopentene alkene only;
iii.   olefin addition through the backbone alkene only;
iv.   oxidative crosslinking by exposure to O_2;
v.   further metathesis reactions induced by residual catalyst.

We report here a full analysis of *f*PDCPD crosslinking and clarify the proposed structure of the resultant material based on solid state $^{13}C\{^1H\}$ NMR, FTIR, Raman spectroscopy, thermal gravimetric analysis (TGA), and differential scanning calorimetry (DSC).

2.2.1 Synthesis and purification of monomer *f*DCPD and polymer *f*PDCPD

The synthesis of the methyl ester functionalized dicyclopentadiene **4** was completed as previously described.[183] The synthesis results in a mixture of two major regioisomers after column chromatography that differ with respect to the position of the methyl ester (i.e. **4** and **3**, Scheme 13). In previous work, we showed that we could selectively polymerize the desired monomer **4** out of this mixture to obtain quite pure linear polymer **7**.[183] For the current work, however, we wanted to ensure that no traces of unwanted monomer **3** participated in the polymerization reaction, since these small impurities might have complicated our subsequent spectral analysis. We therefore removed **3** through conjugate addition of 1,3-diaminopropane to produce the water-soluble derivative **13**.[199e,209] After removal of **13** by aqueous extraction, purified compound **4** was polymerized at room temperature in dichloromethane, using G2 at a loading of 100:1 monomer: catalyst.[183,210]

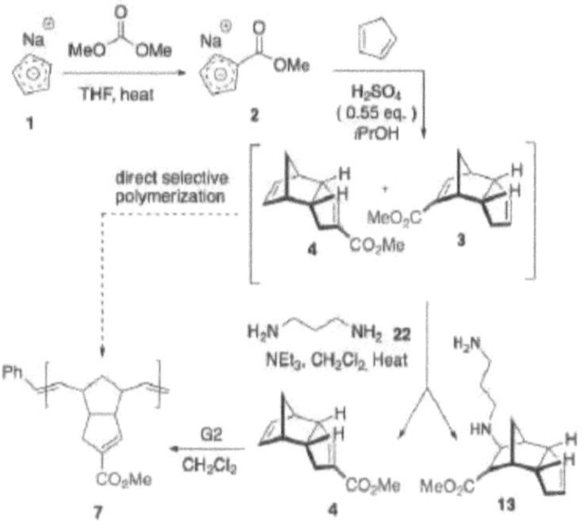

Scheme 13. Synthesis and isolation of functionalized monomer, and metathesis polymerization to afford the desired linear _f_PDCPD.

As reported previously,[194] the product from this polymerization, **7**, is free of any observable crosslinks. It is fully soluble in a range of organic solvents (*vide infra*), and spectral data are consistent with the linear structure shown in Scheme 13. We also explored the polymerization of **4** under neat conditions and at a higher temperature (70 °C). Once again we isolated product that was fully soluble in dichloromethane, and for which spectral data were essentially identical to that of material prepared in solution at room temperature. We can therefore conclude that metathesis-type crosslinks are strongly disfavoured during the initial polymerization of monomer **4**.

After quenching with ethyl vinyl ether, the linear polymer was twice precipitated from dichloromethane and hexanes, which was sufficient to remove any remaining monomer and spent catalyst. Thus purified, polymer **7** (M_n ~20 to 30 kg/mol by NMR integration of the terminal phenyl endcap to the internal olefinic protons) was isolated as an off-white solid but was maintained in solution (CH₂Cl₂) under inert atmosphere at −35 °C to minimize any unwanted oxidation processes prior to the crosslinking experiments.

We initially explored the crosslinking of **7** by exposing it to a variety of elevated temperatures (100 °C, 150 °C, and 200 °C) for 24 hours under an air atmosphere. This resulted in the material becoming insoluble in good organic solvents for the linear polymer (dichloromethane or toluene) which we took as an indication of at least partial crosslinking. A control sample maintained at −35 °C for 24 hours retained solubility in dichloromethane.

All samples were then examined by $^{13}C\{^{1}H\}$ solid-state NMR spectroscopy. As shown in Figure 10A, the carbonyl carbon in **4** was assigned to a singlet at 165 ppm, while the most downfield peak within the broad set of olefinic signals (143 ppm) was assigned to the alkene carbon β to the carbonyl. The remaining signals within this broad multiplet were assigned to the α carbon of the acrylate motif and to the backbone alkenes. Upon heating to increasing temperatures, we observed a consistent reduction of the intensity for the acrylate carbonyl group, accompanied by the emergence of a new broad signal at ~175 ppm. However, even upon heating to 200 °C we did not observe a complete change in the NMR spectrum, indicating that only partial crosslinking had been achieved.

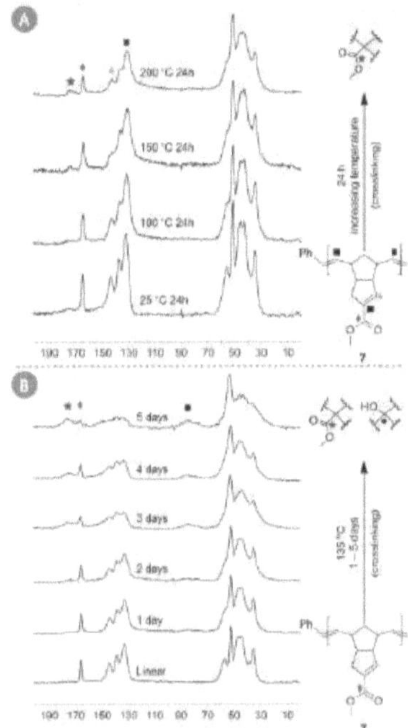

Figure 10. Solid-state $^{13}C\{^{1}H\}$ NMR spectra following thermal crosslinking: A) treatment at 25, 100, 150, and 200 °C over 24 hours; B) treatment at 135 °C over 5 days.

We reasoned that the lack of complete crosslinking at 200 °C might be due to the polymer chains becoming trapped by the first few crosslinks before they could find and react with one another. If this were true, then a slower crosslinking process at a lower temperature (where chain mobility would be more competitive with the rate of thermal crosslinking) should be more effective. To test this hypothesis, a fresh batch of **7** was treated at 135 °C for 1–5 days. As shown in Figure 10B, this led to a much higher crosslinking density, as demonstrated by the nearly complete loss of the singlet at 165 ppm. This was accompanied by the clear formation of the previously described broad peak at 175 ppm. Significantly, this change in chemical shift is almost precisely what one would predict (within a few ppm) for the conversion of a methacrylate-type carbonyl in the starting material to a pivalate-type carbonyl in a product derived from crosslinking through the electron-deficient olefin in the pendent cyclopentene moiety (Figure 11).[211]

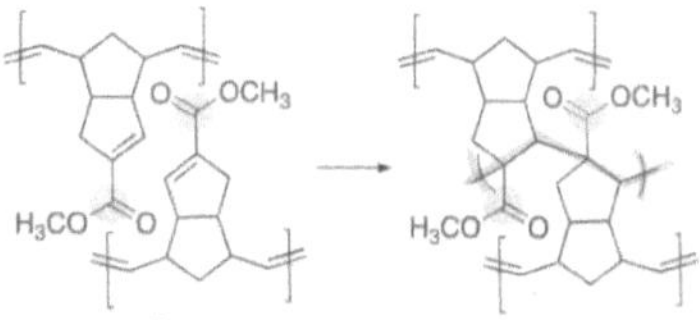

7
Methacrylate-type carbonyl
Predicted: ~167ppm
Observed: ~165ppm

14
Pivalate-type carbonyl
Predicted: ~176ppm
Observed: ~175ppm

Figure 11. Comparison of observed chemical shift data to calculated NMR shifts for a proposed crosslink.

At longer timepoints, we also observed the appearance of a very broad signal centered on ~84 ppm. This would be consistent with the formation of a 3° allylic alcohol motif (as shown in Figure 12) resulting from autoxidation of the alkene polymer backbone. Oxidation of traditional PDCPD has been reported by many other groups[142b,205] and is known to result in incorporation of >30 wt% oxygen with sufficient ageing. Further oxidation of the backbone olefins would result in broadening of the signals corresponding to both the carbinol and alkene carbons and would eventually start to erode the olefinic carbon signals. This is all consistent with the spectroscopic changes shown in Figure 10.

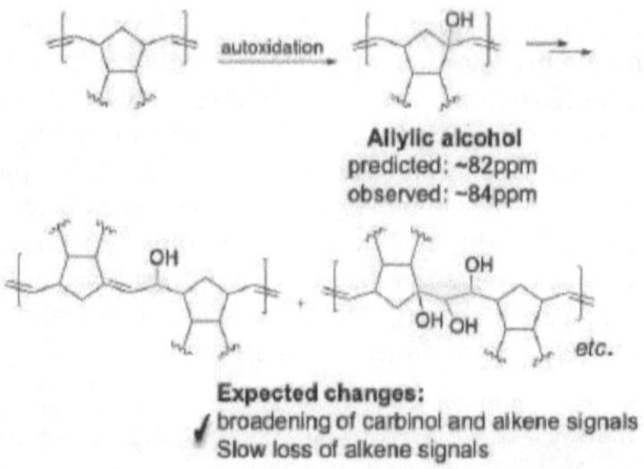

Figure 12. Comparison of observed chemical shift data to the calculated NMR shifts for an autoxidation product.

Mechanistically, autoxidation proceeds through the formation of intermediate radicals,[111b,e] and certainly at longer timescales these would be expected to play a role in crosslinking. Indeed, in our previous report,[194] we described an apparent oxidative crosslinking process that resulted in solutions of *f*PDCPD growing cloudy over a period of days. A concomitant increase in hydrodynamic radius was observed over the same time period.[194] However, the accumulated evidence from the solid-state NMR studies shown in Figure 10 argues against oxidative processes being responsible for the primary crosslinking mechanism. In the short-term heating experiments described in Figure 10A, for example, no carbinol peak was observed, and yet the products resulting from these experiments were clearly crosslinked. Even for the reaction in Figure 10B, crosslinking preceded the appearance of signals arising from oxidation. Heating polymer **7** in the absence of air also results in the formation of product that has all the characteristics of being extensively crosslinked, and TGA/DSC studies under inert atmosphere still show a thermal crosslinking event taking place. Oxidative processes can clearly then be considered a secondary crosslinking mechanism and will not be discussed further here.

By contrast, the changes associated with the carbonyl signal (summarized in Figure 11) strongly implicate the methacrylate group in the primary crosslinking process, and rule out metathesis-type crosslinks as being relevant for our ester-functionalized polydicyclopentadiene. (A ring-opening metathesis process of the type highlighted in purple in Scheme 10A would leave the crosslinked product with an open-chain methacrylate function, for which a second signal at ~165 ppm would be expected.) But these data do not conclusively distinguish between a scenario in which only the acrylate olefin is engaged in polymerization, and one in which the backbone olefins react with the acrylate olefin to establish the molecular crosslink. Integration of the signals in the NMR spectra might have been useful in distinguishing between these two scenarios, but this was not possible with the cross-polarization/magic angle spinning (CP/MAS) experiment that was used to acquire the data.

We therefore turned to FTIR (Figure 13) and Raman spectroscopy (Figure B5-8 in Appendix B and Figure C2-5 in Appendix C) in an attempt to better monitor olefin reactivity and conversion. Interesting changes were observed in the FTIR spectra upon thermal crosslinking (see discussion below) but spectral broadening and peak overlap (as well as fluorescence, in the case of the Raman data) made it difficult to draw any concrete conclusions from the data, without a better way to concretely assign the signals in the spectra.

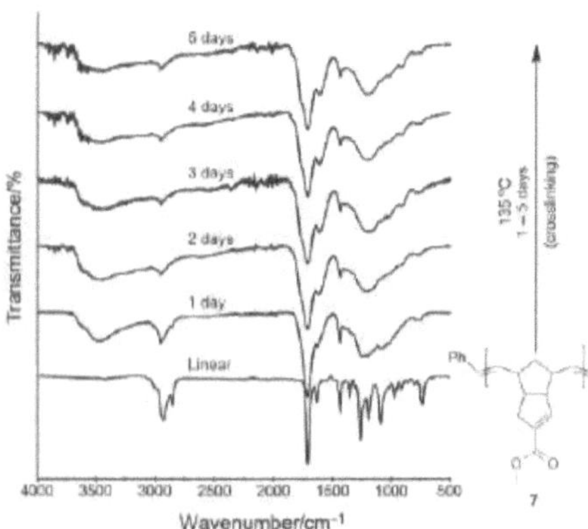

Figure 13. FTIR crosslinking experiments for *f*PDCPD at 135 °C for 1 to 5 days.

Motivated by a desire to better distinguish between the backbone and methacrylate olefins in the vibrational spectra, we next sought to add separate deuterium labels to the norbornene and cyclopentene halves of our substrate. This should shift the alkene C–H stretches from the 3000 cm^{-1} region of the IR or Raman spectra (where they overlap with multiple other signals) down to ~2200 cm^{-1}, where the spectra are unimpeded by competing signals. At the same time, the change from a C–H fragment to a slightly heavier C–D fragment at the ß-position of the methacrylate group should shift the corresponding C=C stretch by about 20 cm^{-1},[212] making it possible to unambiguously assign this signal in the infrared spectrum and possibly follow it through the crosslinking process.

To this end, freshly cracked cyclopentadiene was exhaustively deuterated by the method of Raymond and coworkers,[213] and the resulting Cp-d_6 was used to prepare selectively deuterated monomers **15** and **17**, as shown in Scheme 14. Briefly, this involved reacting either nondeuterated or perdeuterated cyclopentadiene with sodium hydride and dimethylcarbonate to prepare the corresponding carboxylated salt (i.e. **2** or the perdeuterated congener), and then reacting this species with the other cyclopentadiene isotopologue following reprotonation. Removal of the unwanted regioisomer (**3**-d_5 or **3**-d_6) by selective addition of 1,3-diaminopropane (as discussed above) provided the target monomers. Each of these was then separately reacted with G2 to afford the correspondingly deuterated linear polymers **16** and **18**.

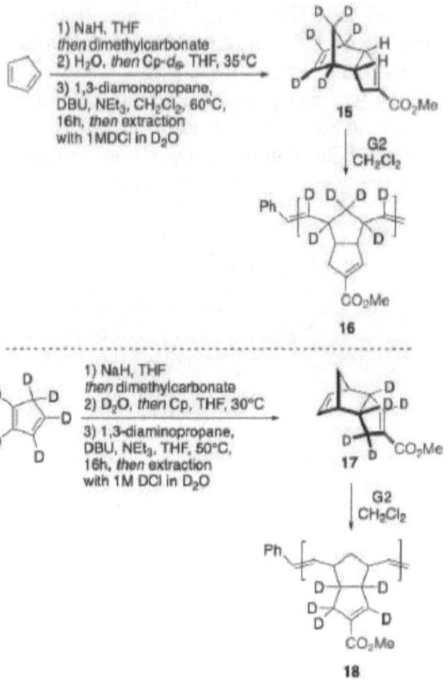

Scheme 14. Synthesis of selectively deuterated monomers and linear polymers.

Spectral data (Figure 14) confirmed the presence of deuterium in the two new polymers. While the C–D stretches were weak in the IR spectra, they were quite apparent (and at least in the case of **18**, relatively well resolved) in the Raman spectra. Moreover, the expected shift in the IR stretch of the methacrylate C=C bond, from 1635 cm^{-1} in polymers **7** and **16**, to 1615 cm^{-1} in polymer **18**, was readily apparent. This allowed for the unambiguous assignment of this critical absorbance.

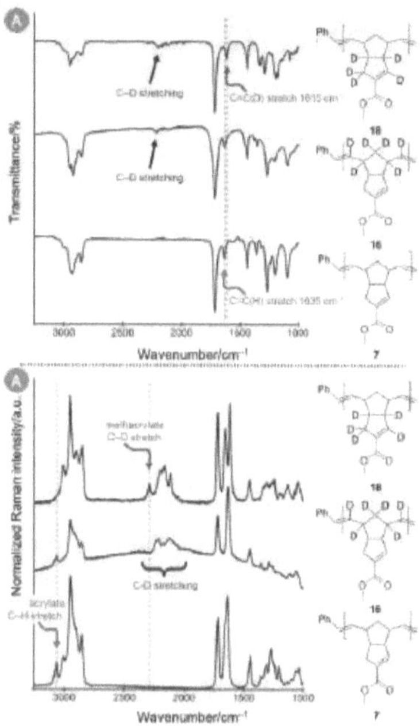

Figure 14. Vibrational spectra for selectively deuterated linear polymers: A) FTIR data; B) Raman data.

The two deuterated polymers were then crosslinked at 135 °C for 5 days, with FTIR and Raman spectra being collected each day (see Figure 15 as well as Figures B5-8 in Appendix B). Broad signals were observed below 1500 cm^{-1} in the IR spectra upon crosslinking, and the very weak C–D stretches at ~2200 cm^{-1} thwarted any effort to directly follow the fate of the backbone olefins. Moreover, extensive signal broadening and fluorescence in the Raman data made it difficult to extract useful information from these data. Nevertheless, important trends could be observed in the carbonyl/alkene region of the FTIR spectra, particularly for polymer **16**. As shown in Figure 15A, the methacrylate C=C(H) stretch at 1635 cm^{-1} clearly diminished upon crosslinking, as did the corresponding methacrylate C=O stretch. At the same time, we observed the appearance of a new carbonyl stretch at higher wavenumbers–consistent with the formation of the pivalate-type carbonyl function that would be expected as a result of olefin addition processes–and a more pronounced (and broad) olefin stretch at lower wavenumbers that would be most consistent with the introduction of greater strain into the backbone alkenes as the pendent methacrylate groups undergo crosslinking.[214] Similar trends in the IR spectra could be observed upon crosslinking of the backbone-deuterated polymer **18** (Figure 15B) and the original undeuterated polymer **7** (Figure 13) but in both these other cases overlapping signals made for a less compelling data set than that observed in Figure 15A.

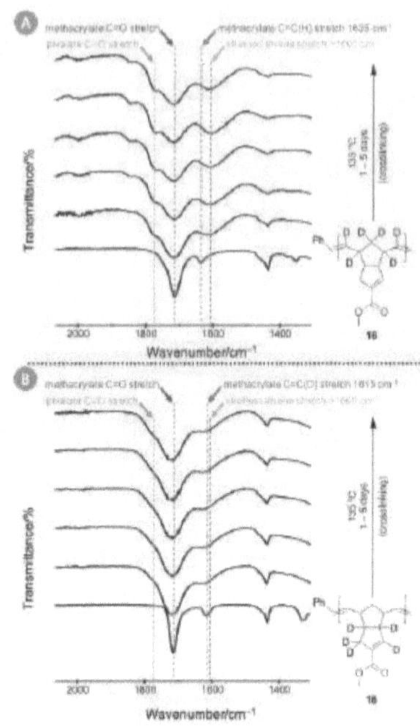

Figure 15. Changes in FTIR data following crosslinking of the two deuterated polymers. Refer to the Figure B5-8 in Appendix B for full spectra.

Together, these details from the FTIR spectra once again implicate addition polymerization through the methacrylate motif as being the most important mechanism of crosslinking for ester-functionalized PDCPD. But yet again, we were unable to completely rule out the participation of the backbone alkenes. In an effort to establish with greater certainty whether or not the backbone alkenes played any role in the (early stage) crosslinking phenomenon, we sought to prepare an analogue of polymer **7**, wherein the methacrylate alkene was selectively hydrogenated.

Preparing the corresponding monomer **19** turned out to be surprisingly challenging, since the cyclopentene motif in compound **4** proved resistant to standard nucleophilic reduction conditions (e.g. Stryker's reagent, dissolving metal reduction, etc.). Eventually, however, we were able to effect selective reduction of **4** to the desired intermediate **19**, using the recently-developed pinacol-borane / diazaphospholene catalyst system (Scheme 15).[215,216] Monomer **19** was then polymerized to the corresponding linear polymer **20** under standard ring-opening metathesis conditions.

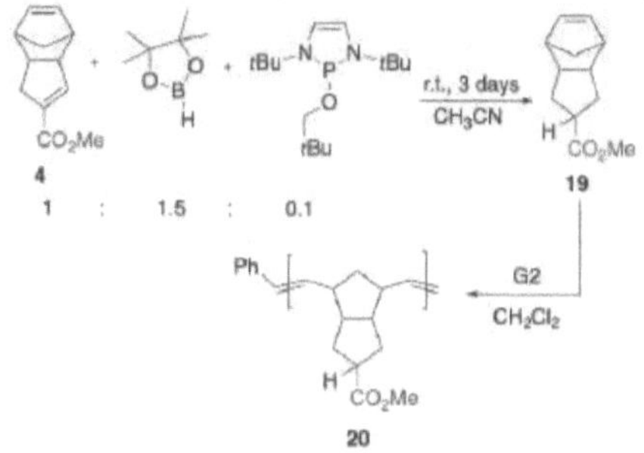

Scheme 15. Synthesis of selectively reduced monomer and polymer.

Polymers **7** and **20** were then examined by TGA and DSC in order to directly observe any crosslinking events and decomposition processes, and to measure any changes in the glass transition temperatures. As we reported previously,[194] polymer **7** exhibits an irreversible exothermic transition at about 145 °C, which corresponds to the onset of crosslinking during the DSC experiment. A steep glass transition was then observed at 173 °C (refer to the solid blue line in Figure 16A). Heating was stopped at 200 °C, and the now-crosslinked sample was allowed to cool to 50 °C before being rerun. We observed no significant change in the T_g for this second run (dashed blue line in Figure 16A), nor did we observe any further evidence of crosslinking. In thermogravimetric analysis experiments (Figure 16B), our functionalized PDCPD behaved similarly to unmodified PDCPD, exhibiting thermal stability to well over 300 °C. Moreover, when the material does eventually decompose it does so through an apparent single-step process (again, analogous to unmodified PDCPD) suggesting that thermal decomposition is not initiated by loss of a labile functional group (as is the case for other functionalized variants of polydicyclopentadiene),[183,184a] but through the backbone olefin linkages.

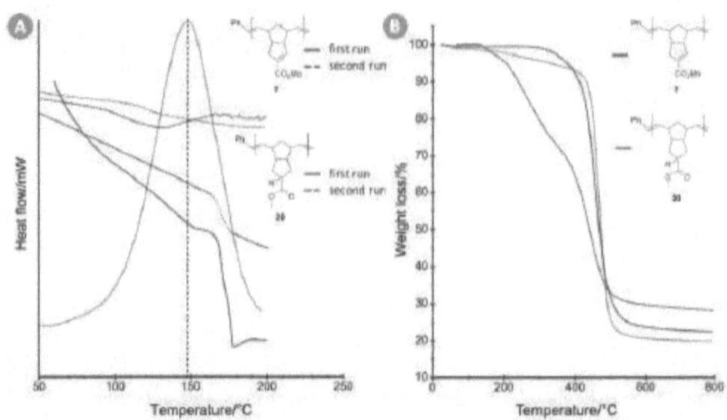

Figure 16. A) DSC thermograms for 1st and 2nd heating cycles of reduced polymer 20 (red) and control polymer 7 (blue). Solid colored lines indicate the first run in each case, while dashed lines indicate the second run. Data shown in grey are for the nonreduced polymer 7 run on a second instrument that is more sensitive to the appearance of the crosslinking exotherm; B) TGA thermograms for reduced polymer 20 (red) and control polymer 7 (blue). Data shown in grey are for a commercial sample of unmodified polydicyclopentadiene polymer.

The reduced polymer **20** behaved quite differently. This material exhibited a much lower T_g (~114 °C; consistent with a mostly linear polydicyclopentadiene), followed by an exotherm at ~130 °C (refer to the solid red line in Figure 16A). This event corresponded to the onset of a decomposition process observed in the TGA experiment (Figure 16B), and likely indicates loss of the more-labile ester (i.e. the methyl isobutyrate function in **20**, relative to the methyl methacrylate function in **7**) though a decarboxylative process. As support for this interpretation, we note that the TGA data show a ~30 % mass loss during this first transition, which agrees very closely to the 30 % of the total mass of **20** that resides within the methyl ester group. After again stopping the DSC experiment at 200 °C and allowing the polymer to cool to 50 °C, we reran the material two additional times. The presumed decarboxylation described above did result in a modest increase in T_g for these subsequent runs (dashed red line in Figure 16A; T_g = 128 °C), but the glass transitions never approached the temperatures expected for an extensively crosslinked polydicyclopentadiene material (>150 °C).

Taken together, our data allow us to draw several important conclusions regarding the nature of crosslinked ester-functionalized polydicyclopentadiene. First, we can largely rule out the existence of any significant degree of crosslinking occurring through ring-opening cross metathesis of the pendent cyclopentene moiety. This is hardly surprising since the presence of the ester function on this alkene should disfavor olefin metathesis at this position. But since metathesis-type crosslinks are so frequently invoked in the polydicyclopentadiene literature (and since they may indeed exist for other Ru-polymerized material) it is important to be able to discount this hypothesis. In the present study, the absence of a second methacrylate peak at ~165 ppm in the solid state ^{13}C{^{1}H}

NMR spectra (not to mention the almost-complete disappearance of the single methacrylate peak that was observed for the linear polymer) allow us to eliminate methathesis crosslinks as being of primary importance.

By contrast, there is abundant evidence in favor of crosslinks occurring *via* olefin addition of the methacrylate alkene, to deliver structures of the form illustrated in Figure 11. Supportive data (Figure A1-7, 34-37 in Appendix A) include the appearance of pivalate-type signals in both the $^{13}C\{^1H\}$ NMR and IR data, as well as the very high glass transition temperature for the crosslinked ester-containing polymer, relative to unmodified polydicyclopentadiene.

Olefin addition reactions occurring through the backbone alkenes were also considered as a possibility, but control experiments with a polymer that contains only backbone alkenes indicated that these were insufficient to produce extensive crosslinking under conditions that we know to achieve crosslinking within the methacrylate-containing material. We can therefore conclude that coupling solely or predominantly through the backbone double bonds is not important for our material.

It is more difficult to prove that the backbone and methacrylate alkenes do not react together to produce mixed mode crosslinks of the type highlighted in green in Scheme 16. However, our DSC data shed additional light on the specific mechanism of the olefin addition process, which allows us to distinguish between mixed mode crosslinks and crosslinks derived solely from coupling of methacrylate functions.

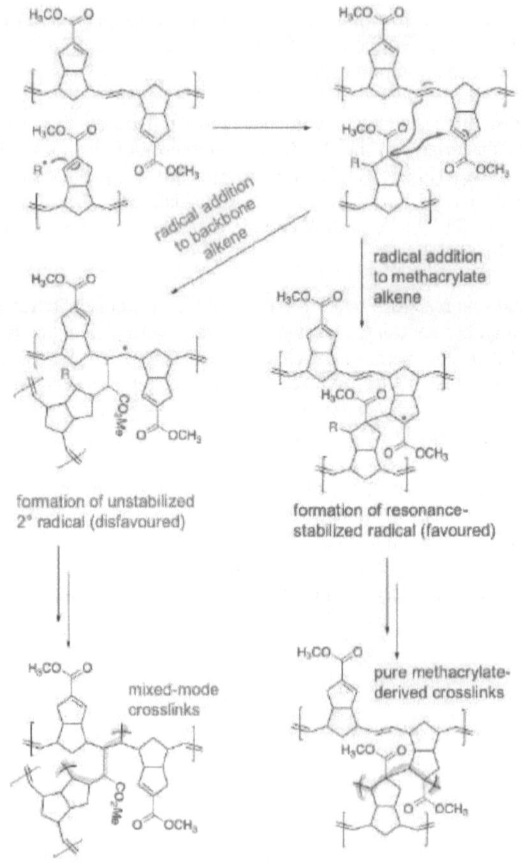

radical addition to backbone alkene

radical addition to methacrylate alkene

formation of unstabilized 2° radical (disfavoured)

formation of resonance-stabilized radical (favoured)

mixed-mode crosslinks

pure methacrylate-derived crosslinks

Scheme 16. Mechanistic considerations favour pure methacrylate-derived crosslinks over mixed-mode crosslinks.

Polymer **7** (containing both electron-poor and electron-neutral alkenes) crosslinks efficiently in the DSC experiment, but polymer **20** (containing only electron-neutral alkenes) does not. This allows us to rule out cation-olefin additions (perhaps initiated by residual Lewis acidic metal catalyst) as being mechanistically significant, since these should proceed equally well through the electron-neutral alkenes of both polymers. We can therefore conclude that thermal crosslinking occurs through self-initiated radical polymerization, as is well known for methyl methacrylate itself.[217]

At each stage of the radical crosslinking process, a radical intermediate will have the opportunity to react with either a methacrylate alkene (blue curved arrows in Scheme 16) or a backbone alkene (green curved arrows in Scheme 16). The former reactions will be much more favored, since the intermediates arising from this process will benefit from stabilization by the carbonyl group.[218-221] Thus, at least at low crosslink densities (before the concentration of free methacrylate groups starts to diminish) we can expect the vast majority of the crosslinks to arise through the addition of one methacrylate group to another.

Finally, we considered whether autoxidative mechanisms played a role in the crosslinking process. Our data indicate that while oxidation (probably at allylic positions) certainly occurs over time, and while this can initiate crosslinking (again, probably through radical additions), oxidative processes are not as significant at short timescales.

2.3.0 Conclusion

Following analysis of crosslinking by TGA, DSC, solid-state NMR, FTIR and Raman spectroscopy, we conclude that the ester-functionalized polydicyclopentadiene introduced previously crosslinks principally through thermal, self-initiated radical coupling of the pendent methyl methacrylate groups. Whereas we could find no evidence to support the existence of any secondary metathesis reactions occurring through the substituted cyclopentene, an abundance of data implicate structures like that illustrated in Figure 11 as underpinning the properties of the crosslinked polymer.

This structural understanding is critically important as we look to exploit *f*PDCPD in different applications. Knowing the chemical structure of the principal crosslinks allows us to rationally design new properties into our material, and will ultimately allow us to take advantage of specific chemical elements to engineer chemically reversible crosslinks in pursuit of a reprocessable form of polydicyclopentadiene.

While we have not explicitly studied the crosslinking of unfunctionalized polydicyclopentadiene in this work, it seems likely that similar olefin-addition reactions[105a,b] are responsible for much of the crosslinking that is attributed to secondary metathesis steps in this parent material, even when ruthenium catalysts are used for polymerization. For example, Lemcoff and co-workers recently showed that exothermic crosslinking events were observed when aged PDCPD samples were analyzed by DSC;[106c] while these data were attributed to the occurrence of metathesis steps following chain relaxation, they would be equally consistent with thermal olefin-addition processes.

Chapter Three: Production and Dynamic Mechanical Analysis of Macro-Scale *f*PDCPD Objects Facilitated by Rational Synthesis and Reaction Injection Molding

The material in this chapter was adapted from "Production and Dynamic Mechanical Analysis of Macro-Scale Functionalized Polydicyclopentadiene Objects Facilitated by Rational Synthesis and Reaction Injection Molding. T. J. Cuthbert, **T. Li** and J. E. Wulff, *ACS Appl. Poly. Mater.*, **2019**, *1*, 2460–2471."

The synthesis of monomer **4** synthesis (half-kilo scale), removal of regioisomer **3**, and polymer **7** synthesis (20 grams scale) were initiated by Dr. Tyler Cuthbert and optimized by **Tong Li**. TGA and DSC sample preparation and measurement were accomplished by Dr. Tyler Cuthbert. The Vickers Hardness and dynamic mechanical analysis samples were prepared by **Tong Li**. The Vickers hardness testing was done by Mr. Ryan Mandau at the University of British Columbia Okanagan. The isolation of monomer **4** and regioisomer **21**, dynamic mechanical analysis (DMA) measurement, reaction injection molding (RIM) protocol development and processing of "UVIC" letters were accomplished by **Tong Li**.

3.1.0 Overview

Incorporation of functional groups into the structure of known polymer materials can provide a means to improve certain desirable properties while maintaining the best features of the existing polymer. At the same time, however, this strategy carries an additional synthetic burden in that the functionalized monomers required for polymerization will themselves need to be made on large scale. For example, while polyvinylchloride (PVC) has certain material advantages over polyethylene that makes it the world's third-most widely produced synthetic polymer[222,223] (despite toxicity concerns[224-226]), the efficient synthesis of the vinyl chloride monomer remains a substantial challenge.[227,228] Currently about 85 % of vinyl chloride is produced through a rather complicated process in which ethylene is first chlorinated to yield 1,2-dichloroethane, which is then thermally cracked to produce the desired vinyl chloride monomer together with an equivalent of HCl.[228,229] The HCl is then reacted with more ethylene in a copper-catalyzed oxychlorination process that generates additional 1,2-dichloroethane for cracking to afford still more of the desired monomer. It is apparent, then, that even for industrially important bulk materials like PVC for which annual worldwide production exceeds 10 billion kilograms each year, synthesis of a functionalized monomer can require considerable attention.

Polydicyclopentadiene (PDCPD) is an example of an industrially important material that could benefit from the incorporation of additional functionality. Produced through ring-opening metathesis polymerization (ROMP) of dicyclopentadiene (DCPD), PDCPD's extensive network of chemical crosslinks provides an extremely high material strength without contributing to excessive brittleness.[110e,153] These properties are maintained at both low temperatures and high temperatures.[11,125,160] After initially being employed to make cowlings for snowmobiles, PDCPD is now used to make body panels, bumpers, and other components for tractors and commercial trucks.[11] Importantly, these commercial products are made using a reaction injection molding process in which the polymerization and crosslinking reactions happen together within the mold to generate the final product.[11,126a,128a,230,231] Spent catalyst remains trapped within the material.

Although useful for many applications, PDCPD has several disadvantages that may be traced back to its lack of functionality. These include a low surface energy when freshly prepared (which makes it difficult to attach PDCPD parts to other objects using conventional adhesives, without first employing a surface oxidation process),[232] a lack of chemical tunability (due to the absence of any functionality other than C–C or C–H bonds), an unpleasant odor (due to residual DCPD being trapped within the final product and then slowly released to the environment)[156,184a,233] and a lack of recyclability (since the bonds formed in the crosslinking steps cannot be undone, as is true of most thermoset polymers).[234] Introduction of a functional group could mitigate many of these disadvantages. While several methods are known to result in functionalization of the residual C=C double bonds within the polymer structure (i.e. post-polymerization functionalization),[142b,149a,181,235] very few options are available for functionalization of the DCPD monomer in such a way that the functional group does not impede polymerization and can be carried through into the polymer product (i.e. pre-polymerization functionalization).[183,184a]

We rationally designed a functionalized form of polydicyclopentadiene (*f*PDCPD, Figure 17), in which an ester group was incorporated at the unstrained olefin within the dicyclopentadiene monomer.[192,194] This material undergoes polymerization and thermal crosslinking similarly to traditional DCPD, with the exception that the two steps can now be done separately. If desired, linear *f*PDCD can easily be isolated and purified prior to the crosslinking event. Although synthesis of linear unfunctionalized PDCPD has been reported by other groups, it is generally more challenging to achieve.[52,88,140,202,236]

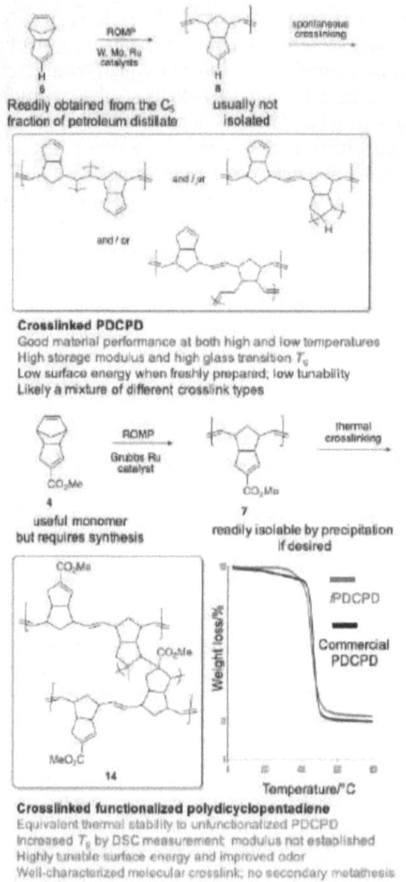

Figure 17. Comparison of traditional PDCPD to *f*PDCPD. Inset shows thermogravimetric analysis for synthesized, crosslinked *f*PDCPD polymer vs. commercial unfunctionalized PDCPD (Product Rescue BVBA).

The incorporation of the ester functional group within our *f*PDCPD confers several advantages—the surface energy of the final product can now be readily tuned through (fractional) saponification, and the monomer itself has a more pleasant smell owing to the presence of the ester.[194] Moreover, while the exact structure of the chemical crosslinks within traditional PDCPD has been a matter of considerable debate,[105a,b] the presence of the ester group reduces the number of reasonable crosslinking reactions. In parallel work, we showed that the structure of the crosslink in our *f*PDCPD is predominantly that shown in Figure 17, in which the critical C–C bond forms through a head–tail linkage of the embedded methyl methacrylate motif.[192] This (at least in principle) provides a means for chemically reversing the thermally formed chemical crosslink, which could in turn form the basis of a recycling process. Critically, the incorporation of the ester functional group does not detract from the desirable properties of PDCPD—in contrast to other known types of functionalized polydicyclopentadiene (wherein the functional group is attached through a labile allylic ester linkage),[183,184a] our *C*-linked ester-functionalized polymer maintains PDCPD's high glass transition temperature, and shows a similar resistance to decomposition at high temperatures.[192,194]

At the same time, our approach begets additional synthetic challenges. Whereas dicyclopentadiene itself is readily available from petrochemical stocks,[237,238] our monomer **4** requires the investment of synthetic effort to attach the ester group. Although an efficient method for the production of **4** was developed based upon our earlier Thiele's ester research,[239] this method unfortunately does not afford **4** as the sole regioisomer. Instead, a mixture of regioisomers is generated, in which **4** is only the second-most abundant species.

We previously showed that flash-column chromatography could be used to isolate a mixture of **4** and its principal regioisomer, **3**. While the ratio of **4**:**3** was unfortunately not tunable through the addition of Lewis acids or other additives, we found that monomer **4** could be selectively polymerized from this mixture to afford the desired linear *f*PDCPD polymer, which could in turn be subjected to thermal crosslinking. Although the protocol efficiently provided both linear and crosslinked polymer for study, it was deemed insufficient for scale-up to production quantities, because of the following reasons:

i.   Separation of the mixture of **4** and **3** from other species employed a chromatographic separation, which would be costly on a large scale;

ii.   Selective polymerization of **4** from the monomer mixture necessitated the separation of the resulting polymer **7** from unreacted **3**; this was typically accomplished through centrifugation, which would be prohibitively expensive on large scale;

iii.   Polymerization from a monomer mixture would be incompatible with reaction injection molding, which is the dominant technique employed for the commercial production of PDCPD automotive parts;

iv.   The synthesis and polymer precipitation made use of flammable solvents, which would be undesirable in an industrial setting;

v.   In our initial work, high catalyst loadings were reported (40:1 substrate:catalyst), which would increase the cost of the final product since the metathesis catalyst used for this work (G2) is relatively expensive;

vi.   Although the mixture of **4** and **3** was prepared on a reasonable scale (up to 45 g), only milligram-quantities of polymer were synthesized initially, raising questions about how scalable the polymerization would be.

In this chapter, we address all of the above challenges and also demonstrate a proof-of-concept reaction injecting molding process for *f*PDCPD. These advances permit the creation of macro-scale *f*PDCPD objects suitable for more advanced materials testing.

3.2.0 Large-scale production of *f*DCPD monomer 4

We began our study by seeking to increase the scale of our initially-developed route to the linear *f*PDCPD polymer, while minimizing the use of flammable solvents (especially the diethyl ether used previously for polymer precipitation) and avoiding the use of chromatography for the isolation of the **4:3** mixture. This necessitated a more thorough assessment of the various species formed through the Diels–Alder reaction of carboxylated cyclopentadiene (i.e. the protonated form of **2**) with cyclopentadiene itself.

As shown in Scheme 17 (and presented in more detail in Chapter Seven Scheme S1), extensive spectroscopic analysis revealed the presence of the four species that had been described previously (heterodimers **4** and **3**, together with homodimers **5** and **6**),[194] along with one additional minor regioisomer **21** visible in the crude NMR spectrum. The connectivity of the newly identified regioisomer–which is consistent with our earlier mechanistic predictions[199d,240] (see Figure E2 in Appendix E)–establishes within the molecule a strained norbornyl C=C double bond without any additional substituents. We were therefore concerned that **21** might participate in the polymerization reaction alongside **4** or, worse yet, react non-productively with the ruthenium catalyst to shut down the reaction. Fortunately, a series of polymerization trials with mixtures of **4**, **3**, **21** and **5** in various states of purity showed **21** to be a harmless spectator under our standard polymerization conditions (room temperature reactions using G2). We found that selective polymerization proceeded efficiently so long as the more-reactive dicyclopentadiene (**5**, formed from the excess cyclopentadiene used in the Diels–Alder step) was removed first (*vide infra*), and provided that only hexane-soluble material was used for the reaction.[241]

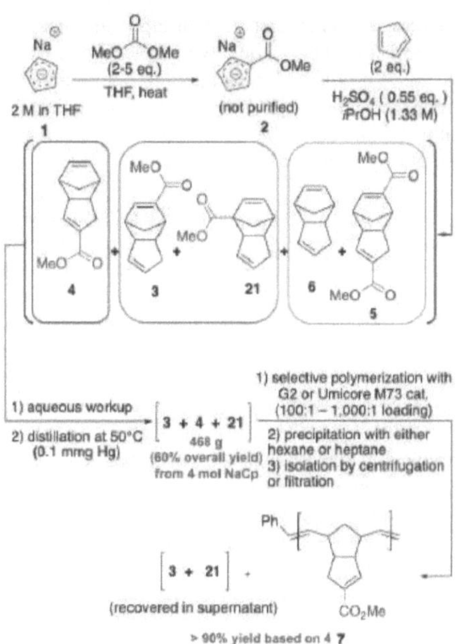

Scheme 17. Improved conditions for producing *f*DCPD and *f*PDCPD on large scale. Compounds 4, 3, 5, 6, and 21 are all produced as endo Diels–Alder adducts.

The critical observation that isomer **4** could be selectively polymerized out of even more complex mixtures than had been envisaged in our earlier work set the stage for the eventual large-scale monomer synthesis described in Scheme 17. Turning our attention first to the carbonylation of sodium cyclopentadienylide (NaCp; **1**) to provide **2**, we found that we could reduce the amount of dimethyl carbonate from 5 equivalents down to 2 equivalents without any significant loss in yield. Perhaps more importantly, we found that even on large scale no additional solvent was required for the carbonylation step; we merely used the THF that the initial NaCp reagent was prepared in.[242] Kilo-scale production of polymer would most likely start from commercially sourced NaCp in THF (e.g. Boulder Scientific sells NaCp as a 20 wt% solution in THF/toluene, in 85-kg cylinders).

In previous work aimed at the synthesis of Thiele's esters, we showed that sodium salt **2** could be purified by washing with diethyl ether.[239] However, in the interest of minimizing the use of flammable solvent, we omitted this step here. Instead, volatile materials (THF and dimethylcarbonate) were simply removed from **2** through rotary evaporation, and the residue was combined with dicyclopentadiene, isopropanol, and 0.55 equivalents of sulfuric acid to promote the controlled re-protonation of the substrate, followed by heterodimerization. Stirring for 48 hours produced the mixture of dicyclopentadienes shown in Scheme 17, in which key intermediate **4** was typically present as about 15–20 %

of the total mixture. A simple aqueous workup using hexanes and water was sufficient to remove inorganic byproducts, along with other impurities that were found in control experiments to cause problems for the upcoming polymerization step. Although it was not necessary to remove **5** (since it does not readily participate in metathesis polymerization), its lower solubility in hexanes relative to the other species in the reaction mixture meant that it was also mostly excluded at this stage.

The use of 2 equivalents of dicyclopentadiene in the Diels–Alder reaction is necessary to favor the formation of heterodimer **4** over the competing production of Thiele's ester **5**. The excess dicyclopentadiene, however, necessarily leads to the production of homodimeric unsubstituted dicyclopentadiene **6** by the end of the reaction period. This can be recycled, provided that it can be efficiently removed from the reaction mixture. Indeed, it is vital to do so, since it would otherwise participate in the polymerization reaction. While copolymers of **6** and **4** are undoubtedly interesting (and are being pursued by us in other work; see Chapter 5) they were not the focus of our current scale-up efforts.

After considerable experimentation, we found that gentle distillation at 50 °C and 0.1 mm Hg was sufficient to remove the unwanted dicyclopentadiene from the reaction mixture.[243] Recovered dicyclopentadiene is reasonably pure (see Figure A12 in Appendix A) and can therefore be cracked and reused in the Diels–Alder step.

By following the optimized protocol described above, we were able to carry through material on up to a 4 mol scale, to achieve a yield of 468 g (60 %, based upon the amount of NaH used to prepare the starting material, **1**) of monomer mixture (containing crude **4** (30 %) **3** (57 %) and **21** (13 %)) in a single batch. Only standard laboratory glassware and rotary evaporators were employed, and no chromatography was used throughout the process. This represents a substantial improvement over our earlier synthetic route.

3.3.0 Selective polymerization of 4 from crude monomer mixture

Because monomer **4** appeared to be the only species within the crude mixture that could be polymerized with the G2 at room temperature, we investigated the production of *f*PDCPD through selective polymerization of the crude material.

We found that polymerization from the unpurified **4:3:21** mixture afforded linear polymer **7** that was spectroscopically identical to that prepared from the column-purified **4:3** mixture used for our earlier studies (Figure 18). No incorporation of any other monomer could be identified.

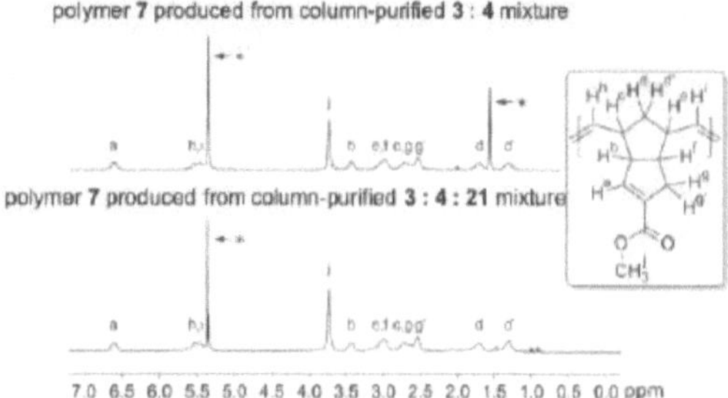

Figure 18. Representative ^{1}H NMR spectra for linear *f*PDCPD polymer prepared from original column-purified monomer[194] mixture vs. that prepared from the unpurified mixture of regioisomers described here. Signals labeled with asterisks indicate CH_2Cl_2 and H_2O in the CD_2Cl_2 NMR solvent. See Figure A21-24 in Appendix A for ^{13}C NMR assignments for the linear polymer, and reference[192] for solid-state ^{13}C spectra of the crosslinked material.

3.3.1 Large-scale production of *f*PDCPD polymer.

We next explored the effect of different catalyst loadings upon both column-purified and unpurified monomer mixtures. In both cases, we observed efficient polymerization reactions with loadings as low as 1000:1 substrate:catalyst—a significant improvement over our initial results.[244] Recognizing that the use of G2 could become problematic on a very large scale (very large scale refers to industrial scale, from hundreds of kilograms to tons; the original Grubbs catalysts are rather carefully controlled by Materia, and are expensive for production scale reactions), we also explored the use of the competing Umicore M73 catalyst (see Figure E1 for the chemical structure of Umicore M73) for which we were quoted a much more competitive price for large quantities. As expected, the Umicore catalyst was also effective at promoting the polymerization of **4** (at a 100:1 substrate:catalyst loading),[244] even from within impure mixtures. For all of these polymerization experiments it was difficult to determine exact rates because with low catalyst loadings the molecular weight of the resulting linear polymer becomes large and solubility is compromised, which complicates NMR analysis (see Table E1 and Figure E4). Suffice it to say, however, that selective polymerization of **4** remained robust for both catalysts across a variety of loadings and starting material purities.

3.3.2 Alternative purification strategies for linear polymer 7

In our initial research, linear *f*PDCPD **7** was precipitated with diethyl ether and collected by centrifugation prior to thermal curing. Both the solvent and the collection method were therefore problematic from the perspective of scale up. We evaluated the use of other solvents for the precipitation step, and found that both hexanes and heptane (a less-flammable and less-toxic solvent that is often used as a process-chemistry replacement for hexanes)[245-248] were effective in precipitating polymer **7** from the mother liquor with no loss in yield.

Filtration was also briefly explored as an alternative to centrifugation. While this facilitated isolation of the linear polymer, benchtop filtration conditions necessarily exposed the product to more air than centrifugation or decanting of the supernatant. It is well known that polydicyclopentadiene is somewhat prone to aerobic oxidation,[144a,183,205] and we have shown previously that 7 undergoes slow oxidative crosslinking when allowed to stand in air.[194] The resulting poor solubility for batches of 7 isolated by filtration led us to abandon this method in favor of simply decanting the supernatant containing the unreactive monomers, in cases where centrifugation was impractical.

3.4.0 Development of a reaction injection molding process for *f*PDCPD

Our efforts so far had greatly improved access to the functionalized dicyclopentadiene monomer 4 in a form that was useful for polymerization, but at this stage the polymer itself still had to be separated from unreacted 3 (as well as smaller amounts of 21) in a tedious centrifugation or decanting step. Moreover, the linear polymer 7 was difficult to form into solid objects suitable for advanced materials testing. In preliminary experiments using a Carver press, we were able to compress 7 into uneven disks that could be thermally cured either in the press or in a separate operation–but objects prepared in this way suffered from defects and voids throughout their macroscopic structure that would have complicated rheology or other measurements. In addition, the use of linear polymer 7 in injection molding operations would require high pressures to flow the polymer into a mold, since the temperature would have to be maintained below the crosslinking onset temperature (ca. 135 °C). For any current industrial process designed around the well-developed reaction injection molding of traditional polydicyclopentadiene to switch over to the use of prepolymer 7 would require a substantial retooling effort that may not be warranted for such an untested material. In order to gain more rapid acceptance of our technology with current users of polydicyclopentadiene, we reasoned that a method where monomer 4 could be directly subjected to reaction injection molding was therefore desirable.

We briefly studied the polymerization of the crude 4:3:21 mixture described above, in the absence of solvent. As expected, the selective polymerization proceeded smoothly, but objects prepared in this way were very soft, and easily deformable (see below for further discussion of this material and accompanying modulus data). Presumably the large amounts of unreacted 3 and 21 that are necessarily incorporated into the final polymer act as plasticizers. In order to develop an efficient injection molding process, we clearly required access to a form of monomer 4 that was relatively free of these contaminating species.

We had already shown that chromatographic separation of 4 and 3 was extremely difficult (although the two compounds are fairly easy to separate from other isomers including 21, they are challenging to separate from one another)—and in any case we hoped to minimize the use of chromatography throughout our processes. Attempted fractional distillation was likewise unsuccessful. We therefore elected to take advantage of the very different reactivity of 4 and 3 toward nucleophiles. It was well known that the strained α,ß-unsaturated ester in 3 was much more electrophilic than the unstrained α,ß-unsaturated ester in 4.[249] Indeed, we[194] and others[249] had previously taken advantage of this difference to purify **4** for characterization purposes. We now sought to test whether a selective conjugate addition could be employed to separate **4** from **3** on preparative scale.

We elected to use 1,3-diaminopropane **22** as the nucleophile in this reaction, reasoning that the product **13** from addition of **22** to **3** would retain a primary amino group and therefore exhibit water solubility, enabling it to be removed from the reaction mixture by a simple aqueous extraction.[250] In the event, diamine **22** proved capable of completely removing **3** from the crude monomer mixture, affording a crude product that was enriched in the desired polymerizable monomer **4** (Scheme 18). After initially optimizing the reaction on small scale in dichloromethane, larger scale reactions (40 g of monomer mixture) were conducted neat.

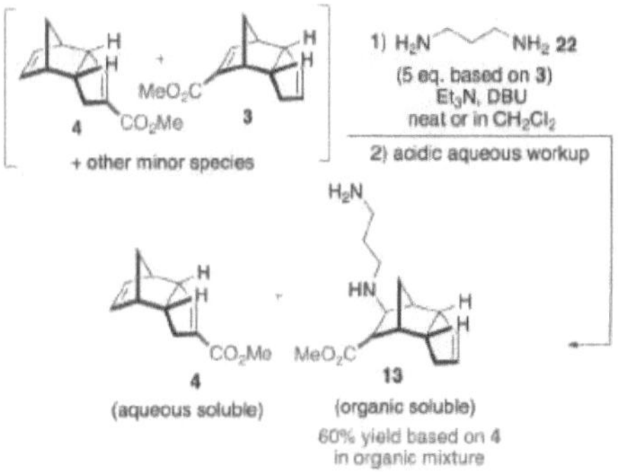

Scheme 18. Selective removal of regioisomer 3 from the crude monomer mixture.

While the conjugate addition described above is sufficient to remove all of the undesired **3** from the reaction mixture (at least so far as can be detected by ^{1}H NMR), the mixture at this stage still contained other impurities, including minor isomer **21** (see below for further discussion and spectral details). In cases where the final monomer, **4**, was purified away from these impurities by column-chromatography, we found that a 60 % overall yield (based upon the concentration of **4** in the original mixture, prior to conjugate addition) of highly purified monomer could be obtained following conjugate addition and chromatography.

With access thus established to large quantities of monomer **4** (with or without the presence of other regioisomers), we next investigated the development of a reaction injection molding process that could be carried out either inside or outside the glove box, and would be compatible with the production of solid objects of suitable dimensions for dynamic mechanical analysis and other measurements.

Commercially available injection molding equipment is generally intended for larger-scale applications than we were interested in here and is most commonly designed for use with thermoplastic polymers rather than thermosets. Even "bench-scale" injection molders therefore have significant dead-volumes associated with single- or twin-screw resin injection (which would lead to considerable wasted sample for us) and lack the ability to heat the mold after delivery of the sample (which is necessary for us to achieve crosslinking). In some senses, reaction injection molding of thermosets is simpler than traditional injection molding of thermoplastics, since monomers can be added to the mold in liquid state and therefore do not require heavy-duty pumps capable of dealing with highly viscous materials. The only caveat is that the mold should be resistant to the temperatures necessary to effect thermal curing.

Recognizing, then, that very simple molding apparatus could be used for our thermoset material, we constructed a mold from a stack of three rectangular pieces of aluminum. Desired shapes were cut into the center piece, while sprue holes were drilled in the top plate to permit addition of liquid monomer and catalyst, the combination of which could be easily added by syringe. The apparatus was easily assembled either in a fume hood or inside of a glove box and could be firmly held together with clamps. Significantly, no special equipment was required either to construct the injection molding assembly or to use it in the laboratory. Pre-cut aluminum plates were ordered as needed from an online vendor, using CAD software to design our desired mold shapes. To highlight the utility of this process for shapes that are more complex than simple rectangular DMA samples, Figure 19 shows a mold used to prepare a version of our University's logo created entirely from *f*PDCPD.

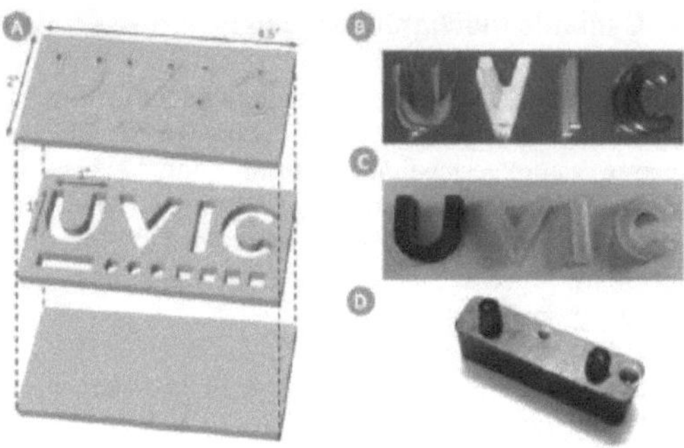

Figure 19. Reaction injection molding apparatus and representative molded *f*PDCPD products. A) A laboratory-scale reaction injection molding apparatus comprised of three aluminum plates. The base plate is solid aluminum. Letters and other shapes have been cut through the center (molding) plate using a CNC mill. Sprue holes are drilled in the top plate to allow delivery of the polymer plus catalyst mixture into the mold. The assembly is clamped during filling, polymerization, and thermal curing (clamps not shown); B) Crosslinked *f*PDCPD objects prepared in silicone molds, photographed on a mirrored surface to illustrate the glassy appearance of the final products. The objects' irregular edges are due to the softness of the silicone molds. This is particularly visible for the letter 'I' which has noticeably curved sides; C) Crosslinked *f*PDCPD objects produced using the aluminum mold described above. Much more consistent dimensions were achieved, although the surfaces have a somewhat roughened finish due to the use of unpolished aluminum for the molds; D) A closeup of a letter 'I' made from crosslinked *f*PDCPD in an aluminum mold. Two small hemispherical voids are visible on the surface, presumably resulting from bubbles trapped within the mold. The cylindrical sprue features are ca. 3 mm in diameter. For all objects, darker colors indicate the use of longer crosslinking times.

Objects prepared using the aluminum mold were much more consistent than those obtained using earlier silicone and PLA molds. While small voids were still observed on the surface of the final products (presumably resulting from bubbles), the overall quality of the samples was sufficient for DMA analysis, in that length, width, and height dimensions were consistent both across individual samples and between sample replicates.

The clamped mold could be easily filled by injection through the sprue holes of a pre-mixed suspension of **4** and an appropriate metathesis catalyst. The polymerization and crosslinking rates for **4** are sufficiently slow that we did not observe any clogging of the syringe needles that we used for injection. Conveniently, the thermal curing step can be accomplished without the need to remove the samples from the mold; the entire apparatus (including *f*PDCPD polymer, aluminum mold, and clamps) was simply transferred to an oven to effect crosslinking. At longer curing times, the samples became darker in color, but otherwise suffered no ill effects.

3.4.1 Dynamic mechanical analysis

With the ability to produce objects of regular dimensions, we next turned our attention to the rheological characterization of *f*PDCPD by dynamic mechanical analysis. In addition to comparing our ester-functionalized polydicyclopentadiene to unmodified dicyclopentadiene, we were curious to know how non-polymerizable impurities within the *f*DCPD monomer, **4**, might affect the performance of objects produced during a reaction injection molding process, where there is no opportunity for unreacted species to be removed from the final product.

As discussed above and illustrated in Figure 20, at different stages during the production of our target monomer we achieved mixtures of **4**, **3**, and **21** in various degrees of purity. Following the initial Diels–Alder dimerization of carboxylated cyclopentadiene (i.e. the protonated form of **2**) with unmodified cyclopentadiene, a mixture of products was observed including Thiele's ester **5** and dicyclopentadiene **6**. However, these two impurities were easy to remove, since the former is relatively insoluble in hexanes while the latter can be distilled out of the product mixture. Following these two operations, the ^{1}H NMR spectrum indicates the presence of **3**, **21** and **4**, but relatively few other significant impurities (Figure 20C). Removal of isomer **3** by conjugate addition with diaminopropane affords a crude product that still contains **21** and **4** (Figure 20B), and final purification by flash-column chromatography provides pure desired product **4** (Figure 20A).

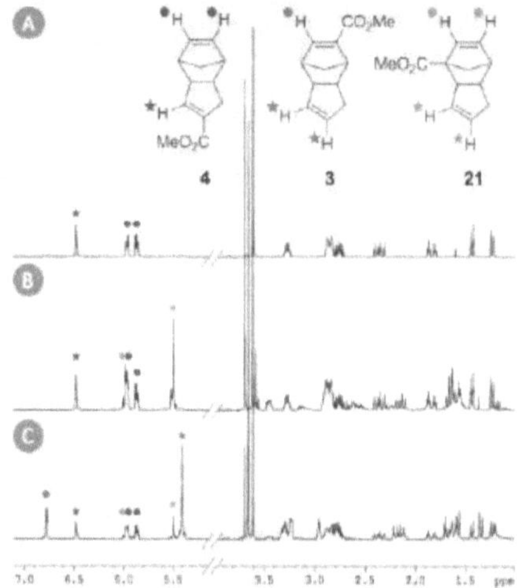

Figure 20. Representative ^{1}H NMR data for monomer 4 at various stages of the production process; A) Column-purified compound 4; B) Product mixture after removal of regioisomer 3 by conjugate addition with diaminopropane, but prior to chromatography. Significant amounts of impurity 21 are present, along with other minor impurities; C) Product mixture after initial formation by Diels–Alder dimerization and subsequent removal of dicyclopentadiene by distillation, but prior to conjugate addition with diaminopropane. Significant amounts of isomers 3 and 21 are both present.

We subjected each of these mixtures to our reaction injection molding protocol, to prepare samples for DMA. Each batch of monomer was polymerized in aluminum molds as described above. The polymerization was carried out in a glove box, using a 1 % loading of the Grubbs second-generation catalyst. Comparator samples of unmodified polydicyclopentadiene were prepared identically. After polymerization, the molds were transferred directly to a 135 °C oven (under air) for either 24 hours or 6 days to effect thermal curing. We previously showed that increased thermal curing across this time regime dramatically increased crosslink density (i.e. decreased the average linear segment length between crosslinks).[192] After thermal curing, samples were removed and analyzed by DMA.

The storage and loss moduli for *f*PDCPD (prepared from column-purified **4**) and unmodified PDCPD were broadly similar (compare Figures 21A and 21C, or Figures 21B and 21D; overlays of storage moduli data shown in Figures 21E and 21F). The prepared *f*PDCPD samples showed a somewhat higher storage modulus at room temperature, but this decreased slightly with elevated temperature, while the unfunctionalized PDCPD samples actually increased slightly in modulus as the temperature was raised. These very minor differences in thermal behavior–which might be attributable to the packing of the ester groups within the polymer lattice–are probably less important than the fact that the data for *f*PDCPD and PDCPD align so closely with one another. Both are very high-modulus materials compared with other organic polymers and exhibit relatively stable moduli across a broad temperature range. At least based upon the DMA data, then, the addition of the ester group does not appear to imperil the physical properties of polydicyclopentadiene. We also conducted Vickers hardness measurements on representative *f*PDCPD and PDCPD samples (Figure 22B). Consistent with the room-temperature modulus data, the *f*PDCPD samples exhibited a slightly higher hardness.

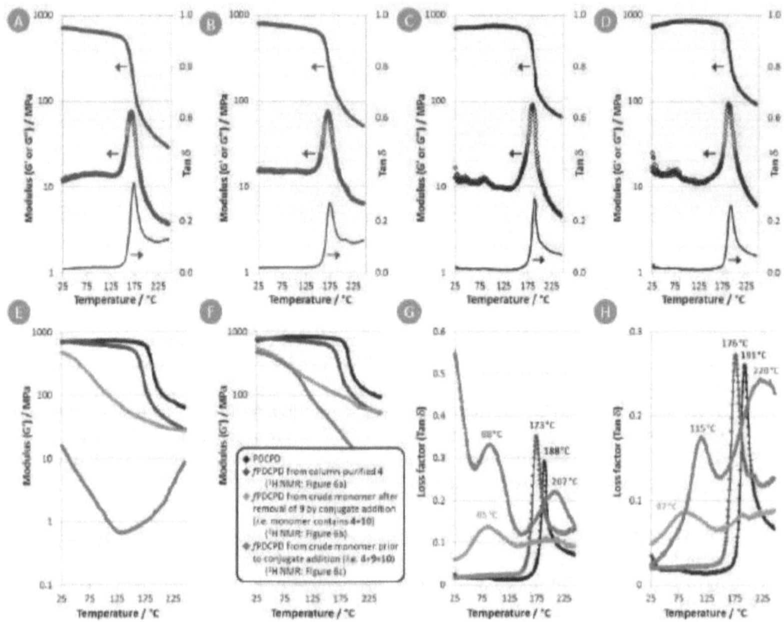

**Figure 21. Results from DMA analysis of *f*PDCPD samples prepared with monomer samples of varying degrees of purity, compared with DMA results for unfunctionalized PDCPD; A) *f*PDCPD from column-purified 4, crosslinked for 24 hours at 135 °C; B) *f*PDCPD from column-purified 4, crosslinked for 6 days at 135 °C; C) Unfunctionalized PDCPD, crosslinked for 24 hours at 135 °C; D) Unfunctionalized PDCPD, crosslinked for 6 days at 135 °C. For panels A–D, data in blue correspond to *f*PDCPD; data in black correspond to unfunctionalized PDCPD; filled data points indicate storage modulus measurements (G'); open data points indicate loss modulus (G''); thin lines indicate loss factor (tan δ); E) Comparison of storage modulus for PDCPD (black data), *f*PDCPD from column-purified 4 (blue data), *f*PDCPD from crude monomer after removal of 3 by conjugate addition (green data), and *f*PDCPD from crude monomer prior to conjugate addition (red data), where each sample was crosslinked for 24 hours at 135 °C; F) Comparison of storage modulus for the four types of samples described in panel E, where each sample was crosslinked for 6 days at 135 °C; G) Comparison of loss factor data for the four types of samples described in panel E, where each sample was crosslinked for 24 hours at 135 °C; H) Comparison of loss factor data for the four types of samples described in panel E, where each sample was crosslinked for 6

days at 135 °C. For panels G and H, numerical values indicate local maxima for each loss factor curve. Samples were run in triplicates.

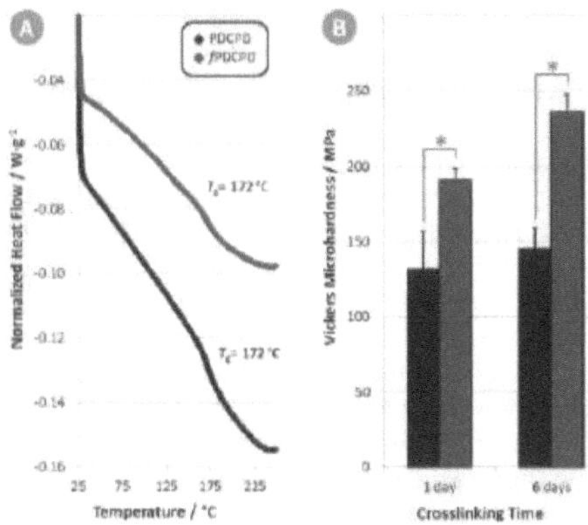

Figure 22. Comparison of DSC and Vickers hardness data for PDCPD (black) vs. ƒPDCPD (blue); A) DSC data showing similar glass transition temperatures for the two samples; B) Vickers hardness measurements showing an increase in hardness for ƒPDCPD relative to unfunctionalized PDCPD, and an increase in hardness with longer crosslinking times. Asterisks indicate statistical significance (p<0.05). Samples were run in triplicates and error bars represent standard deviation. For both sets of measurements, ƒPDCPD was generated from column-purified monomer.

In earlier measurements of the glass transition temperature by DSC (using material prepared under somewhat different polymerization and crosslinking conditions) we found that ƒPDCPD samples displayed a consistently higher T_g (172 ± 3 °C) than that of unmodified PDCPD (155–165 °C).[194] In the present study, however, the DMA data revealed the opposite ordering of the tan δ maxima: as shown in Figures 21G and 21H, the apparent T_g in crosslinked pure ƒPDCPD samples was about 15 °C lower than that for the crosslinked pure PDCPD samples. Glass transition temperature is not a constant for any material of course, and the apparent T_g will change with heating rate or analytical method. To better compare our current samples with those produced in our earlier work, we collected DSC data for samples prepared as described above. As shown in Figure 22A, our ƒPDCPD sample (made from column-purified ester 4) and our in-house prepared PDCPD showed equivalent glass transition temperatures. While this is slightly different than what we had observed previously, the data once again serve to emphasize the similarities in the bulk properties of ƒPDCPD and PDCPD.

Not surprisingly, the modulus of our crosslinked polymer materials decreased with increasing concentrations of impurities in the monomer mixture. As shown in Figure 21E, low-purity ƒPDCPD prepared from the crude mixture of 4, 3 and 21 (prior to conjugate addition with diaminopropane; ^{1}H NMR shown in Figure 20C) and crosslinked for 24 hours exhibited a storage modulus of only 16MPa at room temperature (compared with 706 MPa for fPDCPD prepared from column-purified **4**, and 674 MPa for PDCPD prepared from commercial dicyclopentadiene). Removal of isomer **3** from the crude monomer mixture prior to polymerization and crosslinking (see Figure 20B for ^{1}H NMR of the input monomer, and green curve in Figure 21E for DMA data) greatly increased the modulus of the final product (to 471 MPa at room temperature), despite the fact that compound **21** is still present to act as a plasticizer.

Extended crosslinking times increased the storage modulus of all samples (compare Figure 21E to 21F), but the magnitude of the increase depended upon the provenance of the polymer material. The storage modulus of unmodified PDCPD increased from 674 to 732 MPa, while the same property for pure fPDCPD (i.e. prepared from column-purified **4**) increased similarly, from 706 to 782 MPa). The storage modulus of fPDCPD containing impurity **21** likewise experienced only a modest increase (from 471 MPa to 539 MPa), but the modulus of fPDCPD containing both **21** and **3** increased dramatically upon further crosslinking, from only 16 MPa after 24 hours, up to 470 MPa following a 6-day treatment. Indeed, at temperatures between 60 and 80 °C the moduli for the two impure forms of fPDCPD were virtually indistinguishable after 6 days of crosslinking.

Similar differential increases were observed in the maxima of the loss factor curves (Figures 21G and 21H). While the PDCPD and pure fPDCPD samples experienced only modest increases to the T_g in response to more extensive crosslinking, and while fPDCPD doped with **21** likewise showed very little change upon extended crosslinking, the sample of polymer containing both **3** and **21** (data shown in red) experienced a dramatic change to the loss factor curve with longer thermal curing time.

These data can be rationalized in light of the structures of the various species and what we know about the mechanism of polymerization and crosslinking. As we discussed above, **4** is the only monomer within the mixture capable of undergoing metathesis polymerization at room temperature. After the initial polymerization event, therefore, a polymer sample generated from a mixture of **4**, **3**, and **21** will now contain a mixture of linear polymer **7** together with unreacted **3** and **21** molecules that can function as plasticizers (see Scheme 19). While the ratios of the three species vary somewhat from batch to batch, compound **3** is always the major product. Because of this, samples prepared from the crude monomer mixture–prior to removal of **3** by conjugate addition–will contain very large quantities of plasticizer and will therefore exhibit proportionately low storage moduli when thermal curing times are minimized. As discussed above, these samples are very pliable, and can be easily deformed by finger pressure.

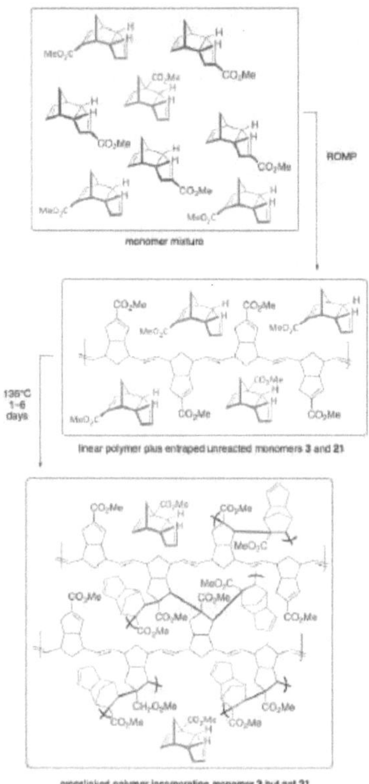

Scheme 19. Proposal for the differential plasticizing effects of compounds 3 and 21 present in the crude monomer mixtures. Polymerizable monomer 4 (or polymer units derived from 4) is shown in blue. Non-polymerizable, but crosslinkable, monomer 4 (or polymer units derived from 3) is shown in red. Non-polymerizable and non-crosslinkable monomer 21 is shown in green. Key polymer crosslinks are indicated with bold black bonds.

We previously showed[192] that *f*PDCPD does not crosslink through secondary olefin metathesis events, but rather *via* head–tail olefin addition polymerization, wherein the methyl methacrylate motif embedded within one residue along the polymer chain adds to the methyl methacrylate motif on a second residue–either on the same chain or (more likely) on a neighboring chain. Critically, compound **3** contains a methacrylate group (an α,ß-conjugated ester) but compound **21** does not! This fundamental difference explains the different behavior of polymer samples prepared with or without regioisomer **3**.

When thermally cured, entrapped compound **3** can participate in the olefin addition crosslinking events, even though it did not participate in the original polymerization. As shown in Scheme 19, then, a polymer that once contained a low crosslink density and large amounts of plasticizer becomes transformed into a highly crosslinked material. A large increase in storage modulus would be expected following this transformation, which is consistent with the observed result.

By contrast, **21** cannot participate in either the olefin metathesis polymerization reaction, or the olefin addition crosslinking process. It likely remains as a spectator (and therefore plasticizer) even after lengthy thermal curing protocols. This explains why the modulus of polymer doped with **21** (green data in Scheme 19) never 'catches up' to that of polymer prepared without this impurity, even at long crosslinking times. It also explains why polymer prepared with both **21** and **3** present (red data in Scheme 19) eventually exhibits very similar modulus values to polymer prepared with only **21** present: after a sufficiently long crosslinking time both samples contain about the same amount of residual plasticizer.

While these explanations necessarily include a certain degree of post hoc rationalization of the experimental data, they nonetheless provide a useful conceptual framework for designing and preparing additional types of functionalized polydicyclopentadiene materials that encompass a broader range of properties (storage modulus, surface energy, T_g, etc.) than has been available in the past.

3.5.0 Conclusion

In this chapter we have made several important contributions toward the broader usage of functionalized forms of dicyclopentadiene (DCPD).

We first demonstrated several improvements to the synthesis of the monomer mixture that leads to *C*-linked ester-functionalized polydicyclopentadiene (*f*PDCPD) and showed that pure linear polymer could be produced from this crude mixture through selective ring-opening metathesis polymerization. The ability to access pure linear (and thus crosslinked) *f*PDCPD without the need for column chromatography represents a significant improvement over our previous work.

Secondly, we showed that our synthetic protocols were capable of producing the functionalized dicyclopentadiene monomer mixture on > 450-gram scale, and we developed a new protocol to remove the principal unwanted species from within this mixture through conjugate addition. This reaction also works on large scale, without additional solvent, and provides product of respectable purity even before chromatography.

Thirdly — and perhaps most significantly — we took advantage of the increased amounts of our materials to develop a reaction injection molding protocol that could be used inside a standard laboratory glove box without the need for specialized equipment.[251] This led to the first production of macro-scale objects from our *C*-linked ester-functionalized PDCPD polymer, which we then used in a series of dynamic mechanical analysis experiments aimed at comparing the mechanical properties of *f*PDCPD to those of traditional unfunctionalized PDCPD. The Lemcoff group has also conducted DMA analysis of their allylically-functionalized PDCPD materials;[184a] as noted above these tended to have lower T_g values as well as lower decomposition temperatures.

While we had previously shown that our vinyl-functionalized *f*PDCPD exhibited nearly identical thermal stability to traditional PDCPD together with an increased T_g (when measured by differential scanning calorimetry), here we showed for the first time that the two polymers are also equivalent in terms of their storage and loss moduli. Given that polydicyclopentadiene is most valued for its excellent material strength, these modulus measurements are arguably the most important piece of data yet in supporting the use of *f*PDCPD in applications that are currently reserved for traditional PDCPD. Indeed, since we have already shown a greatly increased and tunable surface energy associated with *f*PDCPD relative to PDCPD (with γ_{sv} values ranging from 38.5 mN/m up to 66.6 mN/m),[194] we anticipate that this new material can be used in a broader range of applications than is currently open to the unfunctionalized polymer.

Significantly, while chromatographically purified monomer was used to obtain crosslinked polymer with the highest measured storage modulus, we also found that unpurified monomer could still provide a respectably high-modulus material (G' > 500 MPa at 25 °C) without the need for any chromatography steps throughout the entire production process. Thus, while further advances toward the selective synthesis of monomer **4** will no doubt be required to support the large-scale production of the target polymer, the results described herein strongly support the utility of new, functionalized forms of polydicyclopentadiene.

Chapter Four: Harnessing the Surface Chemistry of *f*PDCPD and Exploring Surface Bioactivity

The material in this chapter was adapted from: "Harnessing Surface Chemistry of Methyl Ester Functionalized Polydicyclopentadiene and Exploring Surface Bioactivity. **T. Li**, H. Shumka, T. J. Cuthbert, C. Liu and J. E. Wulff. *RSC Mater*. Adv. 2020, *1*, 1753–1762.[196]"

The *E. coli* growth and antibiotic testing on polymer surfaces, HeLa cell adhesion experiment and fluorescent imaging were accomplished by Hannah Shumka. Preparation of polymer surfaces **23-26**, and the quantification of partially hydrolyzed polymer surface was accomplished by Dr. Tyler Cuthbert. The contact angle measurement was accomplished by Dr. Chang Liu. The synthesis and polymer samples preparation were accomplished by Hannah Shumka, Dr. Tyler Cuthbert, Dr. Chang Liu, and **Tong Li**. Data processing and analysis were carried out by **Tong Li.**

4.1.0 Overview

Polydicyclopentadiene (PDCPD), a polymer produced through ring-opening metathesis polymerization (ROMP) of dicyclopentadiene (DCPD), is used industrially to fabricate vehicle body panels and construction equipment.[2,3,11,22,123,124,125,252] The extensive crosslinking network within PDCPD contributes to high heat and impact resistance,[153,253,254] making it an attractive material for further applications including composite materials,[143b,c,147a] aerogels,[133,138a] and self-healing polymers.[169,175,255] However, PDCPD has a low surface energy when freshly prepared, making it difficult to process or modify, and its lack of chemical functionality—beyond just residual double bonds—limits its broader applications (for examples of post-polymerization functionalization PDCPD through reaction of residual olefins, see[142a,144a,148,181,235]).

To address the low surface energy of PDCPD, and to bestow additional function to the polymer, we installed a C-linked ester group on the dicyclopentadiene monomer[192,194]. The ester-functionalized monomer (*f*DCPD) engages in Ru-catalyzed ROMP similarly to the parent dicyclopentadiene, to afford a linear telechelic polymer that can then be thermally cured to provide a densely crosslinked, functionalized form of polydicyclopentadiene that we call *f*PDCPD (Figure 23) (for alternative approaches to functionalized versions of polydicyclopentadiene, see[183,184a,b]). *f*PDCPD offers a tunable, elevated surface energy, and a higher T_g compared with the parent material.[18] Importantly, both the thermal stability of *f*PDCPD (as assessed by thermogravimetric analysis)[192,194] and the storage modulus (measured by dynamic mechanical analysis)[193] were shown to be similar to those of the parent polymer. The introduction of new functionality while maintaining durability allows for further modifications, potentially increasing the applications of *f*PDCPD for industrial and biomedical use.

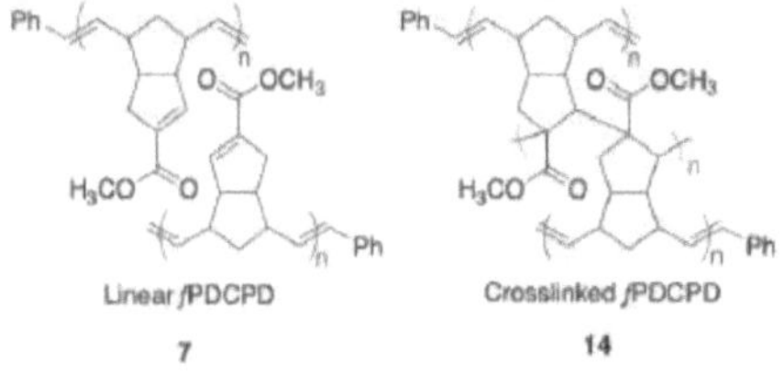

Figure 23. Structure of linear and thermally crosslinked ester functionalized PDCPD.

Through modification of the surface properties of commodity polymers, the response of organisms interacting with the surface can be finely tuned, presenting opportunities to control biological activity.[256-258] From promoting cell adhesion and proliferation,[259] to inhibition of bacterial growth,[260–263] or design of cell patterning surfaces,[264,265] bioactive polymers can find applications ranging from antifouling surfaces[266] to microfluidics[267-269] and tissue engineering.[270-272]

Functionalized surfaces can also be reactive; similar to drug releasing nanoparticles, surfaces may release small molecules or polymers upon reaction or stimulation.[273]

Releasing antibiotics directly from a surface would result in a high local concentration required for inhibition of bacterial growth or bactericidal effect without the requirement of high systemic dosing typical of broad-spectrum antibiotics. Excess and misuse of antibiotics has in the past led to the emergence of resistant pathogens from the introduction of antibiotics into the environment.[274-275] Functionalized surfaces have the potential to mitigate this issue by providing a platform whereby antibiotics can be covalently immobilized to a surface, rendering them biologically inactive, until cleavage occurs when in the presence of the target.[276] *f*PDCPD is an attractive polymer substrate for applications of this type because it can be crosslinked into an insoluble polymer appropriate for surface coatings, yet possesses accessible carboxylic acid functionality that can be exploited to attach antibiotics through readily cleaved ester bonds. Ideally, the surface-bound antibiotic will be inactive until it is released from the polymer, resulting in an off-to-on switch similar to a pro-drug strategy.[276]

As a showcase to illustrate the potential of *f*PDCPD as a direct-dosing polymer support— and more generally to demonstrate the utility of this novel functionalized polymer—we sought to develop methodology that would permit the attachment of chloramphenicol to the *f*PDCPD surface (Scheme 20). Chloramphenicol is a broad spectrum antibiotic that inhibits bacterial protein synthesis by binding to the 50S subunit of bacterial ribosomes.[277,278] It is a widely prescribed antibiotic with a relatively low dosing cost, and continues to be considered an essential medicine by the World Health Organization.[279] Critically, the 1,3-propanediol motif in chloramphenicol is key for its biological activity; if either hydroxyl group is removed or altered, the antibacterial effects of the drug are lost.[280] The primary alcohol within this motif is therefore a prime candidate as a point of attachment to the ester functional group of *f*PDCPD, since immobilization through this group would render chloramphenicol bioinert, while release by promiscuous *E. coli* enzymes[281] would restore antibiotic function.[282] In principle, this approach would result in an autonomous stimuli-responsive surface that is capable of dosing the drug only in the presence of bacteria–thereby limiting the release of excess antibiotics to the environment, and minimizing the potential for creating antibiotic-resistant pathogens.

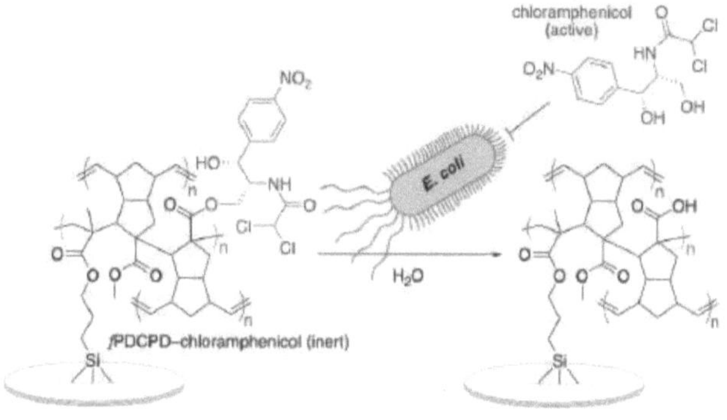

Scheme 20. Attachment of chloramphenicol to ƒPDCPD through an ester bond. When bound in this way the antibiotic has no activity due to blockade of the propane diol motif that is required for biological function. When exposed to *E. coli* the ester linkage can be cleaved by endogenous bacterial enzymes, leading to liberation of the active antibiotic agent.

4.2.0 Harnessing surface chemistry

Understanding the chemical properties of a novel material is paramount to discovering its possible future applications. Although some of the unique properties of ƒPDCPD have been reported previously by our group,[192-194] the degree to which the surface can be further functionalized is not yet completely understood. For example, while previous work has shown that partial hydrolysis of the ester functional groups on the surface of ƒPDCPD can raise the surface energy,[192] we have not yet examined the installation of other functionality at this position. With the goal, therefore, of working toward the conjugation of chloramphenicol to ƒPDCPD, we first needed to gain additional insight into the polymer's ability to accommodate a diverse array of functionality.

To establish that a uniformly functionalized surface could be obtained by adding reagents to functionalized polydicyclopentadiene, we began by pursuing the attachment of a fluorophore to the surface of a ƒPDCPD-coated glass slide. To create the appropriate test-system for this experiment, slides were treated with 3-(trimethoxysilyl)propyl methacrylate and then spin-coated with linear (i.e. uncrosslinked) ƒPDCPD (**7**, Scheme 21) prior to thermal curing. One of the advantages of ƒPDCPD compared with the traditional form of polydicyclopentadiene is that the primary polymerization event (i.e. formation of the linear polymer **7** from monomer **4**) is decoupled from the crosslinking step. This difference allows for a facile preparation of thin polymer samples, since standard spin-coating techniques can be used with the uncrosslinked polymer, which is fully soluble in halogenated solvents.

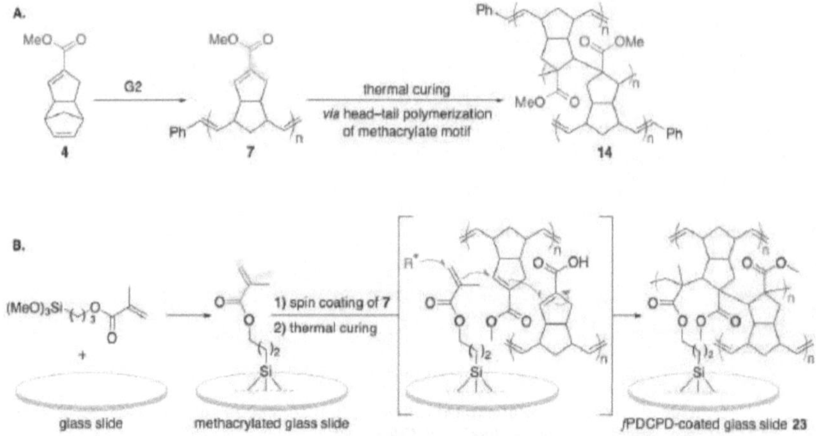

Scheme 21. Harnessing the mechanism of polymer crosslinking to achieve robust surface attachment; A) Structures of the ester-functionalized monomer and linear polymer, and spectroscopically determined structure of the thermally cured material resulting from self-initiated head–tail polymerization across the methyl methacrylate region (highlighted in yellow) within the linear polymer; B) Method for attaching *f*PDCPD to the glass slide, taking advantage of the possibility to spin-coat the soluble linear polymer 7, and to engage the methacrylate motif of the surface activating reagent (highlighted in yellow) directly in the thermal curing event.

4.2.1 Investigation of surface property

In our previous mechanistic studies (see Chapter 2),[192] we confirmed that the principal crosslinking mechanism for **7** does not take place *via* secondary metathesis events of the type that are generally considered for unfunctionalized polydicyclopentadiene[58,59,88,105a,b,187,283-287] but rather through a self-initiated, head–tail radical polymerization event that occurs through the methacrylate motif embedded within **7** (Scheme 21A). In order to achieve robust attachment of *f*PDCPD to glass slides, we further exploited this mechanism by carrying out the thermal curing event on the 3-(trimethoxysilyl)propyl methacrylate-coated slide, with the goal of engaging the methacrylate groups in the surface-activating reagent directly in the self-initiated radical polymerization event that occurs across the methacrylate groups in linear polymer **7**, resulting in the formation of strong covalent bonds that serve to link the *f*PDCPD polymer to the glass slide. While it is impossible to unambiguously prove the adhesive structure **23** indicated in Scheme 21, this method provided much better adhesion of the polymer to the surface of the glass than analogous experiments where **7** was spin-coated and cured directly onto glass slides: in the absence of the 3-(trimethoxysilyl)propyl methacrylate reagent the polymer exhibited a tendency to delaminate from the slide when it was incubated in culture media.

Having thus affixed *f*PDCPD firmly to glass slides, we turned to the conjugation of a fluorophore (Scheme 22). To this end, the samples were submerged in methanolic sodium hydroxide to hydrolyze surface-exposed ester groups, affording **24**. After washing and drying, the functionalized slides were immersed in a mixture of CH_2Cl_2, DMF, and $SOCl_2$ to effect conversion to the corresponding acyl chloride surface **25**. After further washing and drying, the slides were exposed to a solution of 5-TAMRA-PEO3-amine in DMSO containing triethylamine. The presence of the new amide linkage in the target *f*PDCPD–TAMRA conjugate **26** was confirmed by FTIR analysis (see Figure B4, B9-B11 in Appendix B).

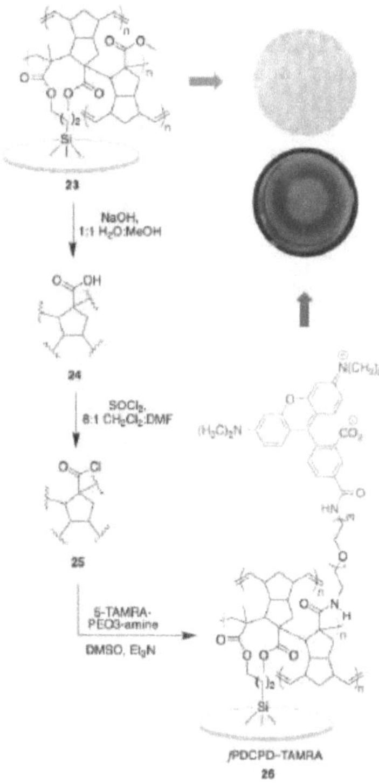

Scheme 22. Construction of a fluorescent *f*PDCPD surface on a glass slide. Inset figures show a visible-light photograph of the initial *f*PDCPD–ester surface (the mottled surface shown in the photo is from a paper towel that was used as backdrop to image the transparent slide) and a fluorescent image of a *f*PDCPD–TAMRA surface acquired using a 576–600 nm excitation and a 610–885 nm emission window. The appearance of red fluorescence across the slide confirms an acceptable level of surface coverage to support subsequent experiments. Refer to Figure S4 for fluorescent images of negative control slides.

To qualitatively evaluate surface coverage, the TAMRA-coupled slides were imaged using a fluorescent microscope. As shown in the inset to Scheme 22, the entire surface exhibited red fluorescence within the limits of optical resolution, indicating a level of surface coverage that could be harnessed to attach other functionality. Control surfaces (e.g. **23** or **24**) showed no significant background fluorescence (Figure S16). The only significant defect appeared to be a series of radial waves in the surface, which result from the spin-coating protocol. Defects on this length scale were not expected to interfere with subsequent experiments. In an attempt to better quantify the density of carboxylate groups available for functionalization, we also treated glass slides covered with **24** with toluidine blue oxide (TBO), a reagent that is known to form a stable complex with surface carboxyl groups.[286] This experiment revealed a surface coverage of up to 0.249 ± 0.125 μmol/cm^2 (see Chapter Seven 7.3.3 and Figure S5 for details), although we found that the result was highly dependent upon crosslinking time. In retrospect, this is not surprising; we previously showed that crosslinking density in ƒPDCPD can be controlled with heating time.[192] More densely crosslinked samples are presumably less able to swell when exposed to TBO in solvent, which results in a lower uptake of the reagent.

To better evaluate our ability to tune the surface of ƒPDCPD through rational functional group manipulations, we next prepared an octyl ester derivative and a tetraethylene glycol derivative and measured the water contact angle on each surface (Figure 24). As expected, the octyl ester derivative **27** was more hydrophobic than the parent methyl ester surface **7**, while the tetraethylene glycol-functionalized surface **28** was more hydrophilic. While the change in contact angle in this experiment was less dramatic than we had previously seen following partial saponification of methyl ester **23**,[184] we nevertheless observed a statistically significant difference between the **27** and **28** polymer surfaces. Together, these data confirm that chemical changes to the ƒPDCPD surface can be used to tune the surface energy in more subtle ways than the hydrolysis protocol used in our earlier work.

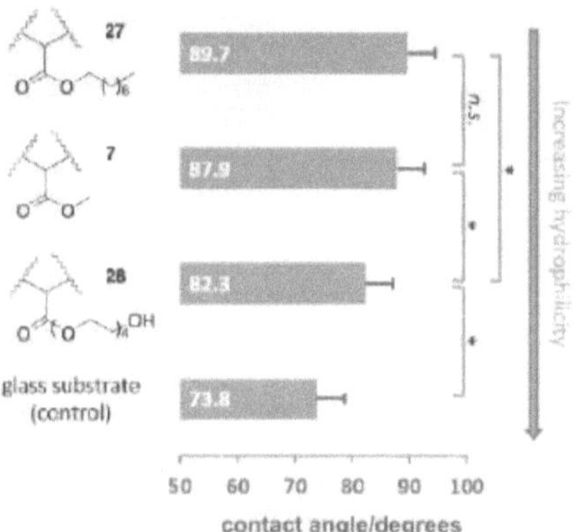

Figure 24. Measurement of water contact angles for three different *f*PDCPD surfaces confirms an ability to controllably tune the surface hydrophobicity. Each value is the average of at least 20 measurements conducted across at least 3 independently prepared samples and error bars represent standard deviation. Statistical comparisons are indicated with a*(p<0.05) or n.s. (not statistically significant).

Moving toward more biologically oriented experiments, and mindful of the utility of controlling polymer–cell interactions for microfluidic or cell-patterning applications, we prepared several *f*PDCPD surfaces containing different functional groups and used these to evaluate the adhesion of tumor cells. Briefly, each polymer surface was prepared on a glass slide that was sized to fit precisely into the bottom of a 24-well tissue culture plate. After inserting the prepared slides into the multiwell plate (using a small amount of epoxy on the underside of each slide to ensure that it adhered to the bottom of the plate), HeLa cells were added in cell culture media. After 48 hours, the surfaces were imaged and then washed with phosphate-buffered saline and then imaged a second time. The populations of adhered cells before and after washing were determined through quantification of the images, and a ratio was taken to determine the amount of cell loss from each surface. As shown in Figure 25, this experiment revealed an impressive ability to control cellular adhesion through surface functionalization. For example, a perfluorooctyl ester derivative was found to be inhibitory toward cellular attachment (90 % loss of cells after washing) relative to the parent methyl ester (43 % loss of cells), while saponification to provide carboxylate groups was found to enhance the degree of cellular adhesion (31 % loss of cells).

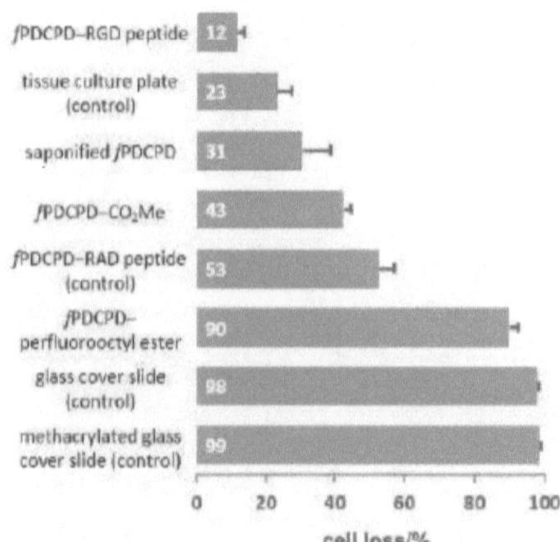

Figure 25. Measurement of HeLa cell loss from a prepared ƒPDCPD surface demonstrates tunable adhesion of tumor cells. Each value is the average of at least 6 independent measurements and error bars represent standard error.

Most significantly, attachment of a cyclic arginine-glycine-aspartate (RGD) tripeptide (which is known to bind to integrin receptors on cells)[259] afforded a surface that was very hospitable for cellular attachment, such that only 12 % loss of HeLa cells was observed following the washing protocol. This was a more effective attachment than the commercial plasma-treated tissue culture plate used as a positive control, which resulted in 23 % loss of HeLa cells after washing. As a negative control for this experiment, we also attached an arginine-alanine-aspartate (RAD) peptide. Despite the fact that the RGD and RAD peptides only differ by a single methyl group (and therefore have similar polarities), the RAD motif has a significantly lower binding affinity for integrin receptors. In the event, use of the RAD-functionalized surface afforded a dramatic reduction in HeLa cell adhesion: 53 % of cells were washed away from the RAD–ƒPDCPD surface, compared with only 12 % from the RGD–ƒPDCPD surface.

4.2.2 Exploring surface biology

Turning at last to the synthesis of the target chloramphenicol conjugate, NMR experiments with the acyl chloride derivative of **4** (as a small-molecule mimic of surface-bound acyl chloride **25**) confirmed that the primary alcohol of chloramphenicol underwent condensation with the acyl chloride in either DMSO-d₆ or CD₃CN, to afford a new ester linkage. We therefore formed the chloramphenicol-functionalized ƒPDCPD shown in Scheme 21 using a similar protocol to that employed for the generation of ƒPDCPD–TAMRA in Scheme 22. Freshly prepared glass slides coated with ƒPDCPD–acyl chloride **25** were submerged in a mixture of triethylamine, chloramphenicol, and DMSO for 48 hours. After rinsing (with both acetonitrile and dichloromethane) and drying, FTIR analysis confirmed the formation of the desired ester bond (Figure B11 in Appendix B).

The slides were epoxied to the base of a 24-well tissue culture plate (as described in the HeLa cell experiments above) and 1 mL of *E. coli* K12 suspension in LB media (at an initial OD_{600} of 0.3) was added to each well. At regular intervals, aliquots were withdrawn to assay the amount of bacteria present. As shown in Figure 26, the *f*PDCPD–chloramphenicol conjugate reduced the growth of the bacteria, relative to controls lacking the chloramphenicol group.

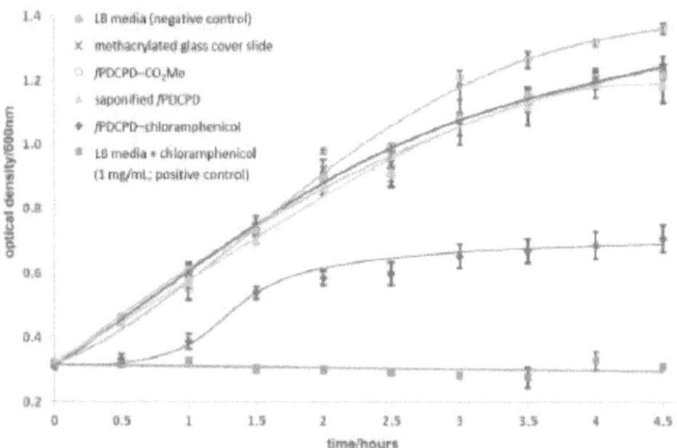

Figure 26. Conjugation of chloramphenicol to *f*PDCPD attached to glass slides creates a bacteriostatic surface that limits the growth of *E. coli*. Each data point is the average of at least 5 independent measurements (9 independent measurements for the *f*PDCPD–chloramphenicol surface) and error bars represent standard error. Lines are meant only to guide the eye and are not intended to convey a precise mathematical relationship.

Chloramphenicol is bacteriostatic to *E. coli* at concentrations lower than 500μg/mL, meaning that it slows or inhibits growth but does not induce cell death.[287] At concentrations greater than 500 μg/mL, on the other hand, bacteriocidal effects begin to dominate, resulting in cell lysis. In a positive control experiment (Figure 26, green squares), 1 mg/mL chloramphenicol was shown to fully inhibit the growth of *E. coli* K12 under identical conditions to those described above (i.e. 1 mL *E. coli* at 0.3 initial OD_{600}, in the same 24-well plate). The use of the methyl ester-functionalized surface (**23**; open circles) or the saponified surface (**24**; orange triangles) resulted in no detectable reduction in bacterial growth relative to the use of either the methacrylated glass slide (red crosses) or the unmodified 24-well plate with no glass slide present (pink circles).

The sigmoidal growth curve obtained for *E. coli* incubated over the *f*PDCPD–chloramphenicol surface (data indicated with blue diamonds in Figure 26) is consistent with the proposed mechanism outlined in Scheme 20. At early time points (< 1 hour), readily available chloramphenicol (perhaps held on the surface through physical interactions rather than through chemical bonds, or perhaps especially susceptible to secreted bacterial esterases) slows, but does not completely eliminate, the growth of the bacteria. At some stage, however, this modest growth rate produces enough bacteria that the organism is able to overwhelm the small amount of chloramphenicol that is present and grow at similar rates to those observed in the negative controls (1–1.5 hours). This increase in bacteria, however, means that more bacterial esterase enzymes are present. These begin to liberate surface-bound chloramphenicol through enzymatic hydrolysis. The liberated chloramphenicol eventually reaches sufficiently high concentrations (>1.5 hours) that it can almost fully retard any further growth of the bacteria. Importantly, this reduction in growth (at about an OD_{600} of 0.6) occurred well below the concentration of bacteria required to limit growth in negative control samples ($OD_{600}{\approx}1.2$) due to crowding or nutrient depletion.

While rigorous proof for the mechanism outlined above is beyond the scope of the current work, the fact that the *f*PDCPD–chloramphenicol conjugate clearly acts to slow the growth of bacteria presents a compelling illustration of the potential utility of rationally installed polydicyclopentadiene surface functionalization.

4.3.0 Conclusion

The addition of functionality to commodity polymers can facilitate the development of new applications. We harnessed the ester group present on a novel functionalized form of dicyclopentadiene to attach a wide range of functional groups. The installation of polar or lipophilic chains to the ester resulted in predictable changes to surface hydrophobicity, while conjugation of a TAMRA dye afforded a red-fluorescent surface that could be used to confirm the extent of surface coverage. Building upon these results, we attached an RGD peptide known to interact with cellular integrin receptors and showed that the resulting *f*PDCPD conjugate led to enhanced adhesion of mammalian cells, relative to a commercial tissue culture plate. Other polymer surfaces — including a control surface incorporating an RAD peptide — led to moderate levels of cell adhesion, while incorporation of a perfluorooctyl ester group almost completely abolished cell attachment. Together, these data indicate an ability to rationally and predictably tune the *f*PDCPD surface to either enhance or reduce interactions with biological organisms. To further demonstrate the utility of the *f*PDCPD polymer, we engineered a self-dosing chloramphenicol conjugate that was capable of releasing antibiotic in the presence of bacteria. As designed, this surface limited the growth of *E. coli*, relative to appropriate controls.

Chapter Five: Copolymers of Functionalized and Non-functionalized Polydicyclopentadiene

The material in this chapter was adapted from: "Copolymers of Functionalized and Nonfunctionalized Polydicyclopentadiene. **T. Li**, and J. E. Wulff. *ACS Appl. Polym. Mater.* **2020**, XXX"

All homopolymers and copolymers were synthesized by **Tong Li**. DSC data were collected by Dr. Tyler Cuthbert. Vickers hardness measurements were performed by Mr. Ryan Mandau. Solid-state NMR spectra were acquired by Dr. Mathew Williams. TGA data were collected by Dr. Wen Zhou. Data analysis and processing were carried out by **Tong Li**.

5.1.0 Overview

Polydicyclopentadiene (PDCPD), a commercially important thermoset, is synthesized through ring opening metathesis polymerization from a strained dicyclopentadiene monomer (**6**, DCPD, Figure 27a), using transition metal catalysts.[55,85,135b,252] Because of its light weight and excellent mechanical properties, PDCPD is used to make automotive body panels and components for engineering applications.[11] In industry, polydicyclopentadiene is produced using a reaction injection molding process, where polymerization and crosslinking take place simultaneously.[127a,289,290]

A lack of chemically modifiable functionality within PDCPD (other than the residual double bonds) limits its use outside of the automotive industry, as does a lack of control over the crosslinking density within the polymer.[183,184a] To address these limitations and broaden the scope of potential PDCPD applications, we reported a *C*-linked methyl ester functionalized polydicyclopentadiene (*f*PDCPD) that enables control over both surface properties (*via* modification of the ester group) and crosslinking density (since the presence of the functional group allows for separation of the polymerization and curing steps).[191-194] Importantly, the addition of the *C*-linked ester did not diminish the thermal stability of the crosslinked polymer, as assessed by thermogravimetric analysis.[192-193]

In previous work, we developed a lab-scale reaction injection molding protocol for *f*PDCPD, and used dynamic mechanical analysis measurements (DMA) to show that the storage modulus of cured *f*PDCPD is equivalent to that of the parent, unfunctionalized polymer.[193] However, the *f*DCPD monomer is somewhat expensive (at least relative to its unfunctionalized congener, which is directly available from petroleum sources) due to the need for synthetic and purification operations during its production. This makes our *f*PDCPD polymer unattractive from an economics standpoint.

In order to bring the benefits of functionalization to polydicyclopentadiene while minimizing added cost, we therefore desired to evaluate the properties of *f*PDCPD-*stat*-PDCPD copolymers.[122e,189] Our hope was to maintain the beneficial mechanical properties of the two homopolymers, while reducing the use of the more expensive monomer and introducing an additional variable for modifying the properties of the final material.

5.2.0 Copolymerization approach

Copolymerization is an effective strategy to combine chemically distinct structures and to develop new materials with designable properties. For example, the Johnson group copolymerized DCPD with monomers that contained silicon linkages (e.g. **37**, Figure 27b). The resulting copolymers had similar mechanical properties to PDCPD homopolymers, but were degradable and recyclable due to the presence of the cleavable siloxane groups.[187] Likewise, the Moore group has shown that copolymerization of dicyclopentadiene with other strained bicyclic alkenes (e.g. **38**) can enhance the rate at which the polymer forms in frontal ring opening polymerization (FROMP) manufacturing, while also altering the material properties of the final product.[122e,184b] In work that is perhaps the most relevant to our own studies, the Lemcoff group recently described copolymers resulting from the combination of two different allylically functionalized DCPD monomers (**9**, X=OH and OPr)—although neither monomer was combined with DCPD itself.[184b] Many of the copolymers made from **9** displayed excellent mechanical properties, although the presence of the labile allylic functional group introduces a potential liability from the perspective of thermal stability.[183,184a]

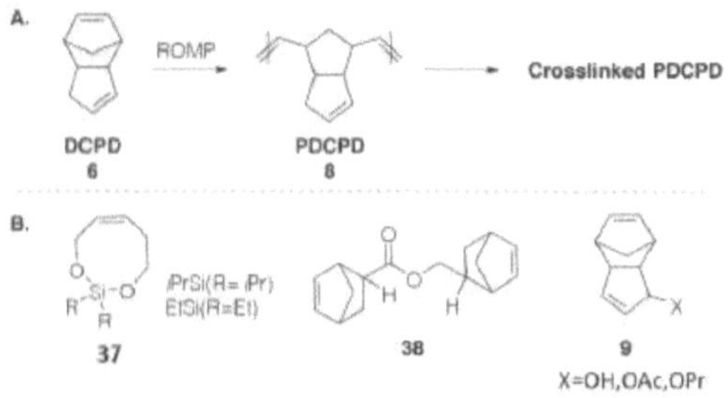

Figure 27. A) Polymerization of DCPD to PDCPD; B) Representative small molecules that have been copolymerized with DCPD or copolymerized with one another. For a full list of comonomers, see Figure E3.[9,87,97,99,109e,101,110a,119,122b,e,127a,183,184a,b,187-189]

In this chapter, we report the synthesis of poly(methyl ester functionalized dicyclopentadiene)-*stat*-(dicyclopentadiene) copolymers (i.e. poly(*f*DCPD-*stat*-DCPD)). Six different monomer fractions were employed, ranging from 100 % DCPD (**6**) to 100% *f*DCPD (**4**, Figure 28). We use solid-state NMR and vibrational spectroscopy techniques to interrogate the structure of the polymers and demonstrate that surface hydrophobicity and mechanical properties including moduli and Vickers hardness can be controlled by balancing the ratio between functionalized and nonfunctionalized monomers. Taken together, these data contribute to an improved understanding of the relationship between molecular structure and material properties within polydicyclopentadiene polymers.

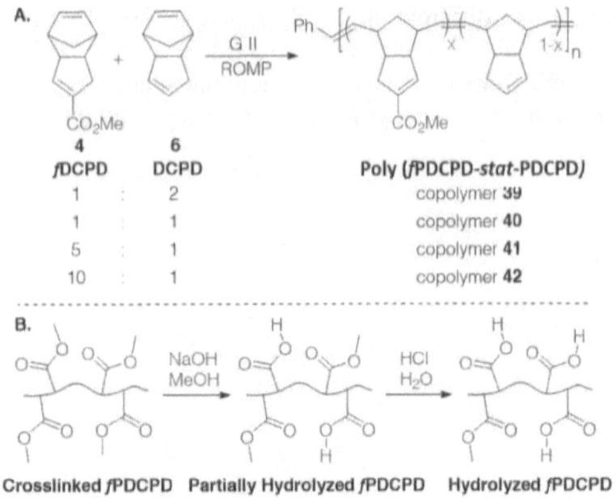

Figure 28. A) Synthesis of poly(*f*DCPD-*stat*-DCPD) copolymers; B) Hydrophobicity change on the *f*PDCPD surface, *via* partial hydrolysis under basic conditions. Solvent-inaccessible groups are unaffected.

Copolymers **39-42** and the PDCPD and *f*PDCPD homopolymers were made by combining non-functionalized and functionalized monomers **6** and **4**[193] in the ratios indicated in Figure 28 and adding 1 mol % of the Grubbs 2nd generation catalyst. This promoter, known for its high functional group tolerance and good activity, is a frequent choice for the production of copolymers.[122e,184b,187] Two different methods were used for polymerization, depending on the experiment that we wanted to conduct with each sample. To provide predominantly linear polymer samples for spectroscopic analysis, the monomers and catalyst were reacted in solvent (dichloromethane) and the product polymers were purified by three rounds of precipitation from dichloromethane and hexanes (refer to Chapter Seven: Experimental Section). While this gave material that was initially fairly soluble and provided spectra that were consistent with a low crosslink density (see below for a discussion of spectral details, and Chapter 2 for a discussion of changes in NMR and IR spectra upon *f*PDCDP crosslinking), solubility was diminished upon exposure to air for even a few minutes, leading in some cases to the production of gels. This is due to a known aerobic oxidation process that is common to both PDCPD and *f*PDCPD,[191] and which provides crosslinks that are structurally distinct from those resulting from thermal curing of *f*PDCPD.[192] The lack of stably soluble material complicated the use of solution state NMR methods or GPC for characterization, and so we relied on solid-state methods instead, as discussed below. To generate samples for analysis of material properties, the two monomers were combined with the Grubbs catalyst in the absence of solvent and were injected into a bench-scale reaction-injection molding (RIM) apparatus[193] inside of a glove box. Once RIM polymerization was complete, samples were cured at 135 °C under air for either 1 hour, 24 hours (1 day), or 144 hours (6 days). Each polymer (for each type of measurement described below) was prepared in triplicate to ensure consistency.

Our previous studies have shown that functionalized monomer **4** polymerizes more slowly than unfunctionalized DCPD (**6**),[193] probably due to intramolecular coordination of the ester group to the ruthenium center in the catalyst.[291] Because of this difference in polymerization rate for the two monomers, it was necessary for us to confirm the incorporation of monomer **4** into the synthesized copolymers prior to evaluating their material properties. We therefore used solid state 13C{1H} NMR, together with FTIR and Raman spectroscopy to confirm the presence of the methacrylate group within the final materials.

5.3.0 Determination of chemical structure
5.3.1 Solid-state NMR spectra

Solid state NMR data (Figure 29) confirmed the presence of both the carbonyl carbon group (165 ppm) and methacrylate alkene (143 ppm) in each synthesized copolymer. The remaining signals visible in the alkene region of the NMR spectra were assigned to backbone alkenes of the polymer, as well as to the pendent cyclopentene groups resulting from incorporation of monomer **6**.

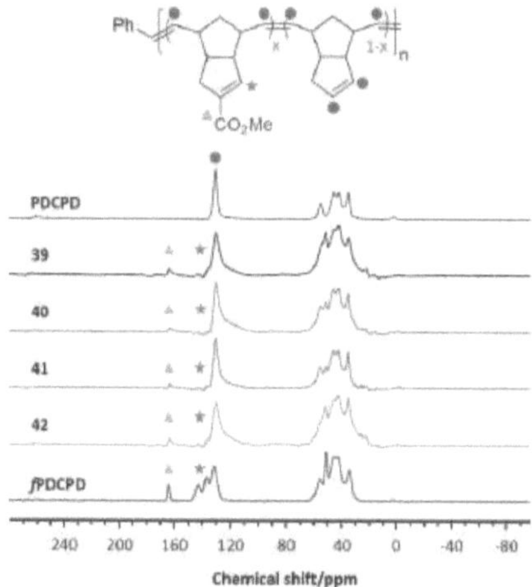

Figure 29. Solid-state 13C{1H} NMR of PDCPD and *ƒPDCPD* homopolymers and poly(*ƒDCPD-stat-*DCPD) copolymers, prior to thermal curing. Three independently prepared batches of each homopolymer and copolymer were made and analyzed. The data presented above come from the third preparation. See Figure A 41 in Appendix A for solid-state NMR spectra of the other two batches.

5.3.2 FTIR and Raman spectrum

In the infrared and Raman spectra (Figure 30), we observed the methacrylate C=C(H) stretch at ~1635 cm^{-1} and the methacrylate C=O stretch at ~1720 cm^{-1}, further confirming the incorporation of the desired functional group. Both of these characteristic stretches were stronger in copolymers **41** and **42** than in **39** and **40**, consistent with the increased amount of monomer **4** used in the production of those samples. Significantly, the absence of a characteristic norbornene stretch in the Raman spectrum (present in the parent monomers at ~1575 cm^{-1}; Figure C5 in Appendix C)[108c,180a] together with the appearance of signals for the unstrained backbone alkene at ~1660 cm^{-1})[111e,f] confirms that polymerization proceeded to completion for all polymers.

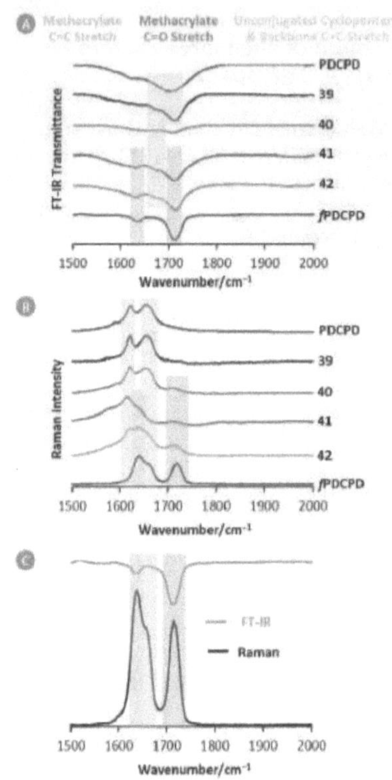

Figure 30. Vibrational spectra for synthesized polymers, prior to thermal curing. A) FTIR spectra; B) Raman spectra; C) FTIR and Raman spectra for linear ƒPDCPD. After thermal curing, background fluorescence precluded the use of Raman spectroscopy for characterization. Three independently prepared batches of each homopolymer and copolymer were made and analyzed. The data presented above come from the third preparation. See Figure B 12 in Appendix B for FTIR and Raman spectra of the other two batches.

5.4.0 Investigation of materials properties

To assess surface energy and hardness, cylindrical samples (14.5mm in height × 14.5 mm in diameter) of all six homo- and copolymers were prepared by reaction injection molding and were thermally cured prior to analysis. When freshly prepared, all six polymers had comparable surface energy, as assessed by water contact angle measurements (Figure 31A). However, upon saponification, those samples containing a higher mole fraction of functionalized monomer exhibited a more substantial decrease in the observed water contact angle—consistent with hydrolysis of the surface ester groups as illustrated in Figure 31B. (Surface oxidation results in some decrease of contact angle even in the absence of any added ester-containing of contact angle even in the absence of any added ester-containing monomer, but those polymers prepared with monomer **4** exhibit additional decreases beyond this background amount.) Interestingly, this decrease in contact angle was not uniform; co-polymer **41** exhibited less of a change than the other three copolymers. This different behavior for **41** was consistent across all three preparations of the copolymers, and presumably indicates a different packing of the polymer chains in **41**, where more of the ester groups are forced inside the bulk of the material and are therefore less accessible to surface hydrolysis. While the cause of this different packing is unclear, it is possible that the different crosslinking mechanisms for ester-functionalized and nonfunctionalized parts of the polymer contribute either to a different crosslinking density or to a different annealed structure within the bulk material, for this specific ratio of monomers.

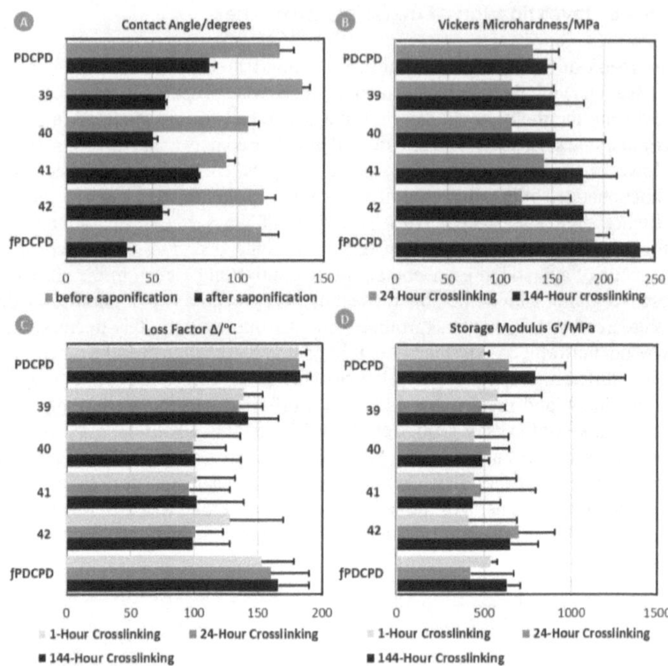

Figure 31. Material properties for synthesized polymers, following thermal curing. A) Water contact angle measured after 1 hour crosslinking; B) Vickers hardness (expressed as modulus); C) Alpha transition; D) Storage modulus at 25 °C. Three independently prepared batches of each homopolymer and copolymer were made and analyzed. The data presented above represent the average result from each measurement. Error bars indicate standard deviation.

A similar non-monotonic trend was observed in the Vickers hardness measurements (Figure 31B).). In these data we saw a consistent increase in hardness (after 24 hours curing) as the mole fraction of the functionalization of monomer was increased—except that **41** once again showed anomalous behavior by displaying a higher hardness than the other three copolymers. Following more extensive thermal curing (6 days instead of 1 day), **41** no longer exhibited such anomalous behavior, and we observed a smooth increase in hardness, moving from unfunctionalized to functionalized polydicyclopentadiene. As expected, all samples displayed increased hardness following longer crosslinking times.

We attempted to measure glass transition temperature (T_g) for all samples by both differential scanning calorimetry (DSC) and dynamic mechanical analysis (DMA). However, while we were able observe clear glass transitions in the DSC measurements for both of our synthesized homopolymers (PDCPD and ƒPDCPD each showed transitions at

172 °C), the glass transitions for the copolymers were ambiguous (see for example Figure D5 in Appendix D).

Turning to DMA measurements (using rectangular samples of 35 mm x 12 mm x 4 mm),[292] we found that all four copolymers displayed lower-temperature glass transitions than the parent homopolymers (Figure 31C), although it is notable that **39** displayed the highest T_g among the copolymers, at ~139 °C, and that the temperature of the glass transition was maintained across all curing conditions. **40–42** all had alpha transitions at around 100 °C (once again, little variability was observed with different curing times). At short curing times (1 hour) we observed two or even three separate peaks in the loss factor data for some samples (Figure D1–3 in appendix D). These secondary transitions mostly disappeared with longer curing times, suggesting that they arise from incomplete crosslinking in the samples that have been heated for shorter periods of time.

We also used DMA analysis to assess the storage modulus of each sample. As we[193] and others[184b] have discussed previously, storage modulus is among the most important properties for any functionalized polydicylopentadiene, since the parent homopolymer finds its greatest commercial use as a structural engineering plastic. Our data (Figure 31D) indicated that all samples maintained high storage moduli, with **39** and **42** outperforming the other two copolymers—although not to a statistically significant extent. The minor differences that we observe in terms of storage modulus could reflect changes in crosslinking density for the different samples, but we were unfortunately unable to accurately measure crosslinking density through swelling experiments, and so are unable to confirm this.

Finally, we collected thermogravimetric analysis (TGA) data for all six cured polymers. We had previously shown that our *f*PDCPD—in contrast to those functionalized polymers derived from allylically-functionalized monomers like **9**—had thermal stability that was comparable to unfunctionalized PDCPD.[192] As expected, the four copolymers behaved similarly (Figure 32), and all six polymers maintained > 80 % of their mass up to temperatures exceeding 400 °C. **39** and **40**, in particular, had identical performance (within experimental error) to the native PDCPD polymer.

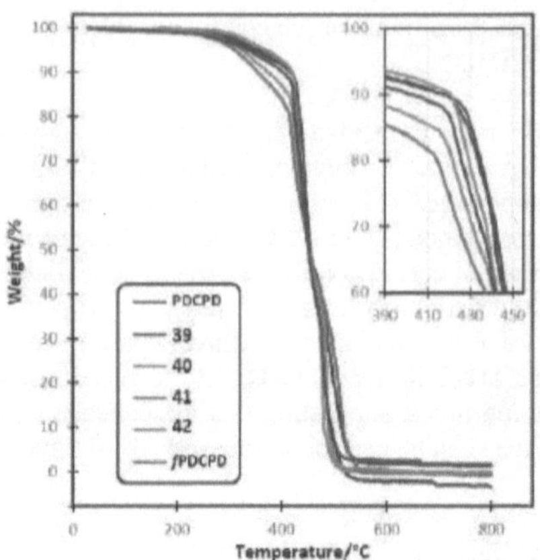

Figure 32. TGA data for thermally cured homopolymers and copolymers. The inset plot shows an expansion of the region of the plot where all six polymers begin decomposing. Samples were in run in triplicates. See Figure D6 in Appendix D for individual TGA plots for each polymer.

5.5.0 Conclusion

In conclusion, we synthesized four different poly (*f*DCPD-*stat*-DCPD) copolymers with tunable surface hydrophobicity, high hardness, high storage modulus, and excellent thermal stability. Random and statistical copolymers are valued in applications ranging from solar cell devices[293,294] and self-healing gels,[295] to wind turbine blades[141c] and eco-friendly biodegradable packaging.[296] The successful synthesis and characterization of the copolymers described here provides important guidance for the future use of functionalized polydicyclopentadiene copolymers in these various applications, where the increased surface energy (relative to the parent homopolymer) could be harnessed for applying adhesives or coatings. Moreover, the presence of the ester group, which we have recently shown can be used as a chemical handle for attaching dyes or drug molecules,[196] suggests additional application for such copolymers in the development of biological materials, for example multifunctional antibacterial catheters[297] or nanocarriers for drug delivery.[298]

One of our principal objectives in undertaking this work was to achieve the benefits that come from incorporation of the ester motif, while minimizing the use of the expensive *f*DCPD monomer (**4**). In this regard, it is notable that copolymer **39**, which was made using only 33 mol% **4**, displayed a surface energy following saponification that was among the highest of all the materials tested, while maintaining a relatively high T_g, high hardness, and high storage modulus—together with a thermal stability that was identical to native PDCPD. This copolymer therefore represents the optimal balance of properties for further exploration. Finally, we note that while the present study made use of relatively high catalyst loadings for operational simplicity, previous work has shown that **4** can be polymerized with loadings as low as 1:1000 catalyst:substrate.[193] While this remains a high initiator concentration by industrial PDCPD standards, it suggests possible improvements to the economics of copolymer production.

Chapter Six: Conclusion and Future Work

"It was the best of times, it was the worst of times, it was the age of wisdom, it was the age of foolishness, it was the epoch of belief, it was the epoch of incredulity, it was the season of the light, it was the season of darkness, it was the spring of hope, it was the winter of despair."

-- Charles Dickens, A Tale of Two Cities.

6.1.0 Overview

A century ago, Hermann Staudinger published his first paper about polymerization, where he described polymers as macromolecules and for which he was awarded the Nobel Prize in Chemistry in 1953. We are now moving to the second century of macromolecular chemistry. As the science journalist Mark Peplow described, polymers are everywhere: the fantastic plastics.[299] These plastics have well-designed, beautiful structures and diverse functionalities. Several generations of polymer scientists have devoted themselves toward the development of novel synthetic routes, the design of 3-D geometries, and the improvement of material properties. Polymers are the materials of 21st century as Brigitte Voit said[300] and I strongly agree. Coming to the end of my Ph.D. studies, I started thinking a lot about what I learned over the past four years and what I have brought to the research group. Surprisingly, I still found myself as passionate about polymers and scientific research as four years ago. Even though there were hard times and tears, I realized how excited I am when I talk about polymers. This book is my contribution to the beginning of the second century of this fantastic field and, even if it is a little step forward in the polymer science world, it is a big step for me as a polymer scientist. I feel the sense of mission but I know that I am a tiny speck in the world of science.

6.2.0 Conclusion

In this book, I have described my contributions—together with those of my coworkers and collaborators—towards understanding the mechanism of *f*PDCPD's crosslinking behaviour, the development of crosslinked *f*PDCPD thermoset materials, and the investigation of its mechanical properties. We aimed to better our understanding of this commercially important material at a molecular level. We expanded the chemistry of PDCPD by attaching a methyl ester functional group and enriched the material properties by adding functionality. By doing so, we broaden the application range of this novel thermosetting polymer.

6.3.0 Future direction of *f*PDCPD

Numerous innovation ideas come from laboratories every year and a large portion of these ideas are successfully transferred into good results and then published in peer-reviewed journals. Materials scientists are synthesizing polymers every day, but how many novel materials have been applied to real markets and have been commercialized? The answer is only a few. Then, how can we apply our fancy plastics in the real world instead of hiding them in a corner of a cabinet or throwing them away after papers get published?

A study from Maine and Seegopaul indicated that commercialization of an innovation idea requires a huge amount of capital as well at least 5-15 years research and development.[301] As shown in Figure 33, science-based businesses have prolonged uncertainty due to changing policies, market fluctuations, and competitors' challenges. The costs of commercialization are way higher than research and development costs.[301] Scientists and entrepreneurs have different mindsets, and often pursue different priorities. In scientific research, the main goals are focused on developing new materials with novel structures and aiming for high impact journals. Professors and researchers have pressure on grant applications and academic achievements. However, R&D in industry is directed toward optimizing material properties to meet customers' needs.

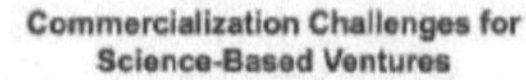
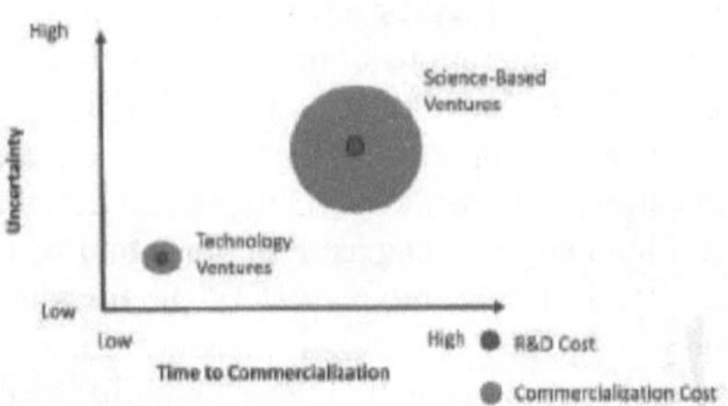

Figure 33. Commercialization challenges for science-based ventures. Adapted from literature.[301]

There is truly a gap between lab-based research and real market applications. It is also a gap of knowledge between scientists and entrepreneurs. Some star scientist entrepreneurs (SSE) have made excellent examples of how to fill the gap. For instance, Robert Langer from MIT cofounded more than 30 science-based ventures. Appropriate strategies were applied including technology-market matching.[302]

An increasing number of university spin-offs, science-based ventures, and university-industry collaborations are emerging. Also, more scientists realized the importance of building up entrepreneurial capabilities.[303] I can imagine a possibility in a near future: scientist entrepreneurs will be more focused on translating technology/science discovery into real markets needs. This might be a new revolutionary change.

I am lucky to enroll in a mini-MBA program in the Beedie School of Business at Simon Fraser University this September. From what I learned from this program so far, the gap between lab-based research and real market applications can be filled if more Ph.D. students become involved in this kind of program and practice their entrepreneurial capabilities at the early stage of their research career. I also believe that there should be more communication between industry and academia. I hope that I can contribute to the commercialization of *f*PDCPD in the future.

6.3.1 Commercialization of the first generation of *f*PDCPD

The synthesis, characterization, and applications of first generation of functionalized polydicyclopentadiene have all been examined extensively in this book. In thinking about the future, it is time to recall the important industrial value of polydicyclopentadiene. I would like to share my innovation ideas about the commercialization of *f*PDCPD.

Human society is facing an environmental and energy crisis. Many efforts have been made to fight it and one in particular is the transition from fossil fuels to electricity in personal vehicles. Battery electrical vehicles (BEVs) on the market right now have already outperformed many of the traditional internal combustion engine (ICE) vehicles. However, the range, i.e., how far the vehicle can run on a single change, is the greatest downside of BVEs. Among all the strategies to improve the range, cutting the weight of vehicles is proven to be an effective way.

Owing to its excellent mechanical properties (i.e. hardness, stiffness, and impact strength), anti-corrosion, and good heat resistance, while maintaining light weight and low manufacturing cost, *f*PDCPD is an ideal replacement of traditional materials (metals) used to produce the body panels and other components in vehicles. This novel material has great potential to change the BVE market by cutting the weight and extending the range significantly.

*f*PDCPD, a thermoset polymer, can in principle be processed through reaction injection molding (RIM) technique at an industrial large scale. Notwithstanding the advances in scale described in Chapter 3 of this book, however, the production of the *f*PDCPD polymer has only been carried out on decagram scale. Although the synthesis of *f*PDCPD has been substantially improved and the fundamental studies of RIM processing conditions were thoroughly carried out (chapter three), challenges still remain and collaborations with industry have not yet furthered commercial development. At the moment, the development of *f*PDCPD has approached a new stage where commercial applications can be envisioned.

6.3.2 Potential applications of *f*PDCPD

Owing to the additional functionality, *f*PDCPD has a wide range of potential applications from the automotive industry and aerospace engineering, to biomaterials and nanomaterials.

Since the protocol of scaling up *f*PDCPD was developed and optimized, it is now feasible to produce large quantities of *f*PDCPD for commercial applications. At the same time, however, the current synthetic route passes through a Diels-Alder reaction that necessarily leads to the competing production of two unwanted regioisomers as well as two unwanted homo-dimeric adducts. While these unwanted products could in principle be cracked to regenerate starting materials, and while we were successful in developing methods to separate most of these undesirable species from the *f*DCPD monomer (**4**) without the need for chromatography, all of this processing still raises the effective cost of our monomer. By comparison, the unfunctionalized DCPD monomer literally comes out of the ground, and can therefore be sourced inexpensively ($0.80 / lb crude; $1.40 / lb in an upgraded form suitable for purification).

One option to gain the benefits of functionalization while minimizing the costs of the final polymer would be to add small amounts of **4** into the manufacture of reaction injection molded DPCDP parts. We confirmed that copolymers made from DCPD and *f*DCPD exhibit similar levels of robust mechanical properties, thermal stability, and tunable hydrophobicity (chapter 5). This leads us in the direction for producing functionalized materials in an economic way. Given the fact that homopolymeric *f*PDCPD and copolymers are both available from RIM processes and that the industrial production of *f*PDCPD follows similar methods for traditional non-functionalized PDCPD manufacture, this copolymer strategy represents the "lowest hanging fruit" for commercial utilization of our functionalized monomer.

We also explored the surface chemistry of our novel material, by harnessing the embedded functional group in order to exert finer control over hydrophobicity, and to control interactions with biological organisms through the conjugation of biologically relevant functional groups. Therefore, *f*PDCPD — or, more likely, less-expensive copolymers—can potentially be applied in nanotechnology for drug delivery nanocarriers, thin film substrates for bio-sensor devices. It is also possible to apply *f*PDCPD in bulk biomaterials like antibacterial catheters, coatings, and adhesives. Other possible applications are seashore pipelines in marine engineering, self-healable rocket frame and turbine blades in aerospace engineering.

6.3.3 The Development of RIM Technique

We realized that the next revolution of RIM techniques requires the development of better mold materials. In collaboration with a group at Oak Ridge National Labs, we replaced the traditional aluminum molds in our RIM apparatus with 3-D printable materials. We selected ultem—a family of polyetherimide (PEI) materials—on account of its excellent thermal stability, anti-corrosion, and light weight. We 3-D printed several well-designed and delicate molds (Figure 34 shows a car shape ultem mold) and produced *f*PDCPD within the molds. We believe the next generation of RIM mold material will be more flexible in terms of 3-D design and materials selection.

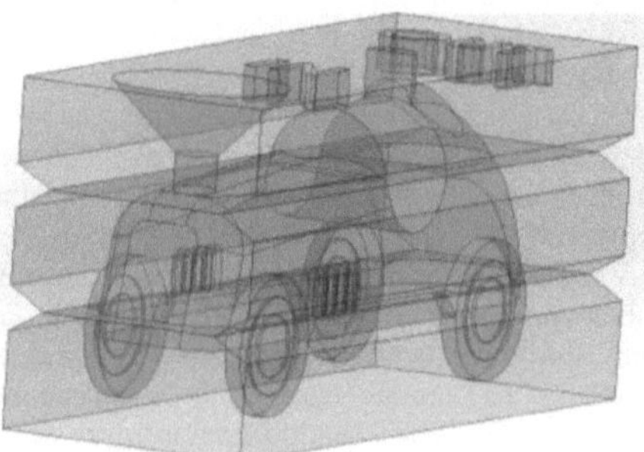

Figure 34. Sketch of 3-D printed ultem mold in a car shape.

6.4.0 Second Generation of *f*PDCPD

Taken together, the data in this book show that *f*PDCPD is a useful material that combines the benefits of the existing commercial polydicyclopentadiene (high strength and toughness, RIM processing) with the added benefits that come from functionalization (tunable surface energy, opportunities for attaching dyes or drug molecules). However, our data also show that the synthesis is far from ideal. As described above, the Diels-Alder reaction that is used for the production of the *f*PDCPD monomer is not sufficiently regioselective. An additional challenge is that the reaction that we currently use to form our monomer—like many Diels-Alder reactions—is highly *endo*-selective. This means that we have so far been limited to the use of *endo*-**4**.

While we have not rigorously measured the polymerization rate of **4**, we have found that ROMP reactions using this monomer require higher catalyst loadings than those employed for unfunctionalized DCPD, and generally proceed more slowly. This is likely due—at least in part—to coordination of the ester functional group to the metal centre required for catalysis.[291] An *exo*-functionalized DCPD monomer would likely react much faster under ROMP conditions (Figure 35). Indeed, even for unfunctionalized DCPD, we know from the literature that the *exo*-isomer reacts up to 20 times faster.[91,122a]

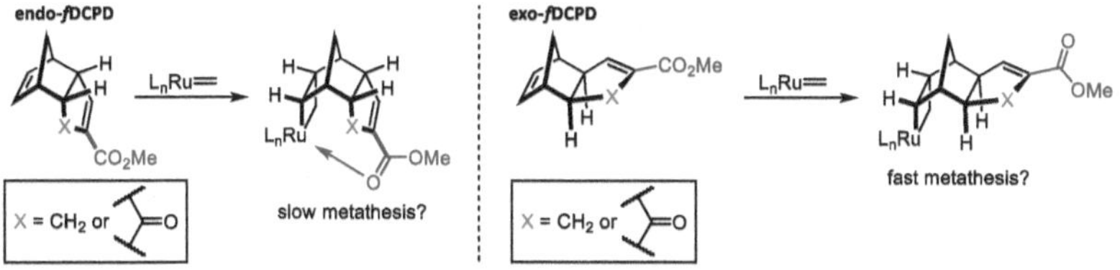

Figure 35. Chemical structure of *endo*- and *exo-f*DCPD monomers and implications for catalysis.

The analysis above tells us that it is time to finally abandon the Diels-Alder reaction that led the Wulff group into this research in the first place,[194,199a,b,239,240] in favour of a strategy that allows for the synthesis of a functionalized dicyclopentadiene monomer in a way that is (1) highly regioselective, and (2) has at least the possibility to deliver the *exo* diastereomer. Ideally the synthesis would also be short and would leverage catalytic reactions for key transformations, so as to minimize cost.

It would be great if we could somehow add CO_2 and methanol to dicyclopentadiene, in the presence of an appropriate catalyst, and produce our *f*DPCD monomer in a form that is suitable for polymerization. However, this is probably not possible—in other known carbonylation reactions of DPCD, the reaction always occurs at the strained alkene in preference to the unstrained alkene where we want to install our functional group (Figure 36).[305,306]

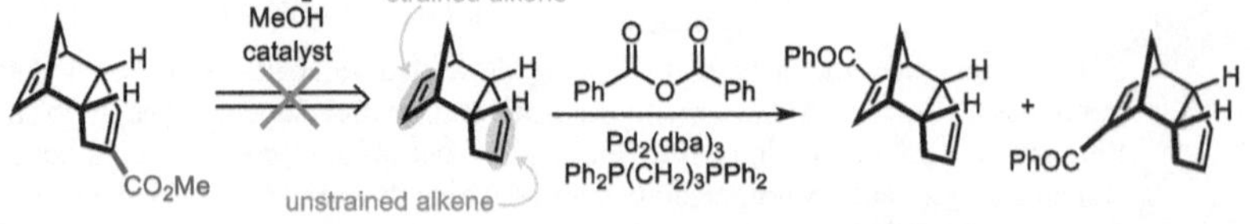

Figure 36. Chemical structure of *endo*- and *exo-f*DCPD monomers and implications for catalysis.

The solution to this challenge comes with redesigning the functionalized monomer itself—keeping the ester group that permits functionalization of the final polymer, and keeping the strong sp^2–sp^2 linkage between the monomer and the functional group that is responsible for the outstanding thermal stability observed in our *f*PDCPD, but adding a ketone at the allylic position (Figure 37). This ketone can be added to PDCPD regioselectively through photo-oxidation in the presence of very low concentrations of a porphyrin catalyst,[307a,b,c,d] and its presence will allow for the easy functionalization of the adjacent vinyl group under catalytic conditions.[308a,b,c] Moreover, this sequence of three catalytic operations should in principle work for either *endo*- or *exo*-dicyclopentadiene,[309a,b,c,d] meaning that the final functionalized monomer should be achievable in either diastereomeric form.

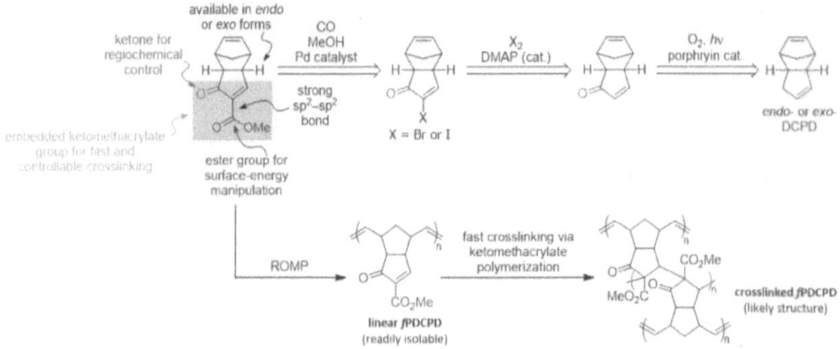

Figure 37. Second-generation *f*PDCPD: design, synthetic route, and likely polymerization/crosslinking pathway.

Far from being a just a synthetic handle, the presence of the ketone also transforms the embedded methyl methacrylate motif in our first-generation linear polymer (**7**) into a ketomethacrylate reminiscent of the cyanoacrylate function in super glue! This will likewise block the formation of secondary metathesis events, allowing for the isolation of the linear polymer if desired (for example to remove ruthenium impurities) but will participate in a lower-barrier head-to-tail radical or anionic polymerization than our first-generation methyl methacrylate motif. This will make for a faster crosslinking (curing) step, thereby improving the overall reaction injection molding protocol.

This second-generation functionalized polydicyclopentadiene will be tackled by a new graduate student in the Wulff group, Ben Godwin.

Chapter Seven: Experimental Section

7.1.0 General experimental methods

Solvents and reagents were purchased from Sigma Aldrich, (St. Louis, MO, USA) and used as received with the following exceptions. Grubbs catalyst 2nd generation (G2) was purchased from Chem-Impex. RGD and RAD cyclic peptides were purchased from Peptides International. 5-TAMRA-PEO3-amine was purchased (as the TFA salt) from ChemoMetec. RPMI1640 media, fetal bovine serum and trypsin–EDTA were purchased from Thermo Fisher. *E. coli* K12 was purchased from ATCC. Chloramphenicol and all other reagents were sourced from Millipore Sigma. Matsunami round glass cover slides and 24-well plasma-treated cell culture plates were both purchased from VWR. Tetrahydrofuran was dried by distillation over sodium and benzophenone. Dichloromethane was degassed by three freeze-pump-thaw cycles and stored over activated 4Å molecular sieves in a nitrogen filled glovebox. Hexane was dried by passage through alumina in a commercial solvent purification system. Triethylamine was dried with CaH$_2$ and distilled under N$_2$. Liquid reagents were transferred *via* a rubber-free plastic syringe with a stainless-steel needle. Organic solvents were concentrated between 35-50°C by rotary evaporation under vacuum. All reactions were performed in single-neck, flame-dried, round-bottom flasks fitted with rubber septa under a positive pressure of argon, unless otherwise noted. All polymerizations were completed in a LC Technologies Inc. glovebox with a N$_2$ atmosphere, and only exposed to air during crosslinking or analysis. Cp-d$_6$ was synthesized according to a literature procedure.[213] Analytical thin-layer chromatography (TLC was performed using aluminum plates pre-coated with silica gel (0.20 mm, 60 Å pore-size, 230-400 nm mesh, Macherey-Nagel) impregnated with a fluorescent indicator (254 nm). TLC plates were visualized by exposure to ultraviolet light (254 nm). Flash-column chromatography was performed over silica gel 60 (Caledon, 63-200 μM).

7.2.0 Instrumentation methods
7.2.1 Nuclear magnetic resonance spectroscopy (NMR)

Solution NMR was conducted on a 300 or 500 MHz on a Bruker AVANCE 300 spectrometer equipped with a 5 mm PABBO BB-1H/D Z-GRD probe at ambient temperature (298–300 K). ^{1}H, ^{2}H, and ^{13}C NMR spectra were referenced relative to the residual solvent peak (CHCl$_3$: ^{1}H δ=7.26, ^{13}C δ=77.0; CH$_2$Cl$_2$: ^{1}H δ=5.32). ^{1}H, ^{2}H and ^{13}C chemical shifts (δ) are reported in parts-per-million (ppm) relative to tetramethylsilane. NMR data are presented as follows: chemical shift, multiplicity (s = singlet, d = doublet, t = triplet, br = broad, m = multiplet), coupling constants (J, reported in Hz), integration in parts per million. Solid-state ^{13}C{^{1}H} NMR spectroscopy was carried out on a 400 MHz spectrometer using a CP-MAS pulse sequence at ambient temperature (298–300 K).

7.2.2 Fourier-transform infrared spectroscopy (FTIR)

FTIR was completed on a Perkin Elmer FTIR Spectrum Two Spectrometer (Waltham, MA, USA) in the universal attenuated total reflectance mode (UATR), using a diamond crystal as well as the UATR sampling accessory (part number L1050231). IR wavenumbers (λ) are reported in units of cm^{-1}.

7.2.3 Raman spectroscopy

Raman spectroscopy was completed on Renishaw's inVia Raman microscopes (Wotton-under-Edge, Gloucestershire, UK) in extended mode with a HeNe laser 633 nm, calibrated with Si wafer (part number:INVIA 1216-10). Crosslinked samples were photo-bleached to minimize fluorescence background. Origin software was used for baseline subtraction, smoothing and final data plotting.

7.2.4 Thermogravimetry (TGA)

TGA was completed on a Q600 SDT from TA Instruments or SHIMADU TGA-50 instrument and analyzed using TA Universal Analysis and Microsoft Excel, under an N_2 at mosphere at a heating rate of 10 °C/min or 5 °C/min up to 900 °C using a ceramic pan with approximately 20 mg of sample.

7.2.5 Differential scanning calorimetry (DSC)

DSC was completed on a DSC Q20 or Q2000 TA Instruments simultaneous thermal analyzer with samples being placed in an aluminum Tzero pan, referenced against an empty aluminum Tzero pan. Data were collected with a ramp rate of 5 °C/min following temperature equalization at 50 °C under a nitrogen atmosphere flowing at 100 mL/min with 5-10 mg of sample.

7.2.6 Gel permeation chromatography (GPC)

GPC measurements were performed using a Viscotek Model 302 liquid chromatography system (Viscotek GPCmax + TDA 302 triple detector array) equipped with refractive index (RI), low-angle light scattering (θ=7 °), and right-angle light scattering (θ=90 °) detection. DCM was used as the mobile phase at a flow rate of 1mL/min, and the column temperature was set at 30 °C. All polymer solutions were filtered through membrane filters with a nominal pore size of 0.45 μm prior to injection into the GPC columns. The data were collected and analyzed using appropriate GPC software from Viscotek. The system was installed with a Tosoh Biosciences, LLC TSKgel HHR series guard and two separation columns in series; specifically, HHR-M guard column and two GMHHR-M columns, respectively. Molecular weights were calculated from GPC data using an algorithm from Viscotek. The dn/dc value for polymer 7 (0.114 mL/g) was calculated from an on line calibration of varying concentrations from 0.1 mg/mL to 1 mg/mL of freshly prepared polymer. The dn/dc value for polystyrene in DCM (0.162 mL/g) was obtained from PolyAnalytik. GPC chromatogram of linear *f*PDCPD 7 is presented in Figure S1.

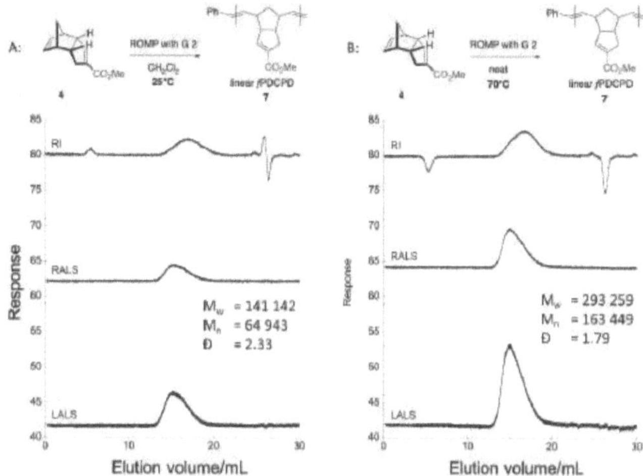

Figure S1. GPC data for polymer 7 prepared under different conditions.

7.2.7 Dynamic mechanical analysis (DMA)

Samples of pure functionalized or nonfunctionalized polydicyclopentadiene were prepared as described below, and tested by dynamic mechanical thermal analysis (DMTA) on an Anton Paar MCR 302 rheometer with SRF 12 geometry and CTD 600 oven. Frequency and amplitude sweeps were completed to ensure testing parameters for temperature-sweep measurements were within the viscoelastic region (Figure S2). Temperature-sweep measurements were performed with a strain of 0.1 %, a frequency of 1 Hz, and a heating rate of 1 °C/min. Tested samples were made as section 7.4.2 described.

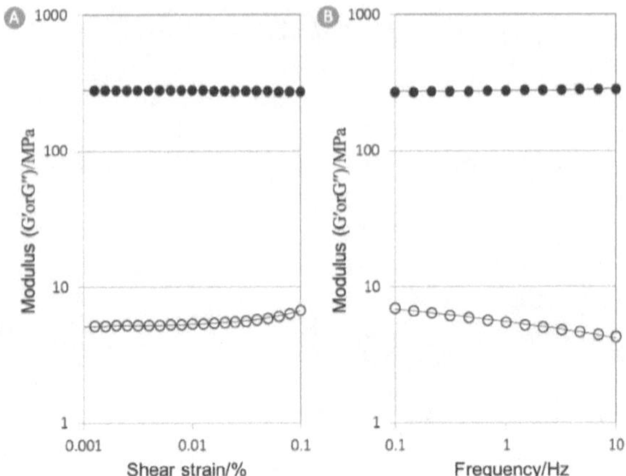

Figure S 2. Amplitude and frequency sweeps for *f*PDCPD from column-purified 4, confirming that testing parameters for the DMTA experiments were within the viscoelastic region of the material. A) Amplitude sweep; B) Frequency sweep. Filled data points indicate storage modulus measurements (G′); open data points indicate loss modulus (G″).

7.2.8 Vickers hardness test

Complementary hardness tests were completed on a Buehler Wilson VH 3100 instrument with a Vickers diamond shaped indenter. Tested samples were made as section 7.4.2 described.

7.2.9 Contact angle measurements

Deionized water droplets were deposited on the flat face of the cylindrical samples described above, using a Hamilton microsyringe with a mechanical dispenser. Water contact angle measurements were carried out with a Holmarc contact angle meter (HO-IAD-CAM-01). Side-view images of the drop on the substrate were taken by a high-performance aberration-corrected imaging lens with precise manual focus adjustment (CMOS sensor). Three water droplets were deposited at three different regions of each surface, and three different preparations of each polymer were tested. A mean contact angle and standard deviation were then determined from the resulting measurements.

Polymer thin film surfaces

Spin-coated polymer thin films were prepared as Table S2 in 7.3.1 section. Samples were thermally cured as section 7.4.2 described. Polymer surfaces affixed to glass cover slides (prepared as described above) were first rinsed with diethyl ether and methanol, then dried *in vacuo*. 3 µL of deionized water was deposited on the surface of the film, and the drop was imaged using a Holmarc contact angle meter. Left and right contact angles were obtained using the software that was packaged with the instrument. Each analysis was repeated on at least three independently prepared polymer samples, and each sample was interrogated at multiple locations on its surface. A total of at least 10 right and 10 left contact angle measurements were averaged to generate the results (Figure 24 and Figure S3).

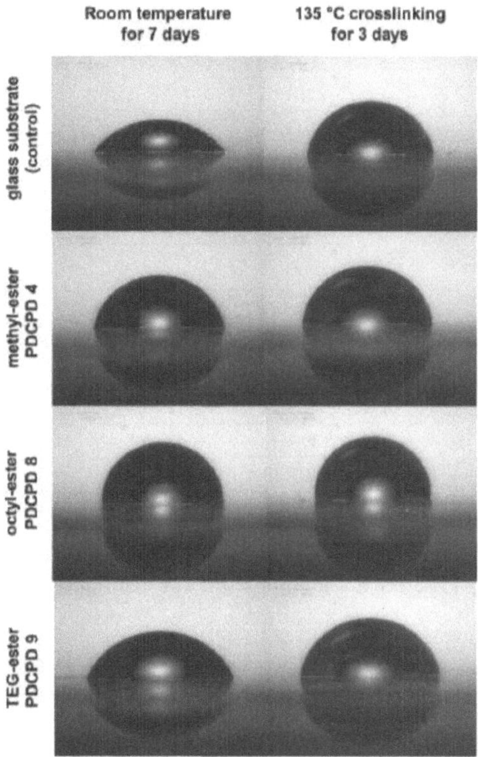

Figure S 3. Representative of contact angle photos.

Cylindrical copolymer surfaces

Tested samples were made as section 7.4.2 described. Samples were put into an oven for 60 min after polymerization to remove any volatile residue, and washed with methanol and deionized water. Samples were then submerged into a 5 wt% solution of NaOH in 1:1 deionized water:methanol for 2 hours and then washed with 1 M HCl. The samples were then thoroughly washed with deionized water followed by methanol, and dried in vacuo prior to water contact angle measurements.

Table S 1. Raw contact angle data for synthesized homopolymers and copolymers prior to thermal curing. Samples were analyzed in triplicate, using three different preparations. SD represents standard deviation.

	Contact angle/degrees	
	before saponification	SD
*f*PDCPD	113.6	10.22
42(10:1)	115.3	6.74
41(5:1)	93.5	5.02
40(1:1)	105.9	6.57
39(1:2)	137.3	4.42
PDCPD	124.3	8.26
	after saponification	SD
*f*PDCPD	35.4	4.13
42(10:1)	56.3	3.59
41(5:1)	77.3	0.5
40(1:1)	50.4	3.09
39(1:2)	57.6	1.25
PDCPD	83.5	4.08

7.3.0 Biological experiments and analysis methods
7.3.1 Fluorescent imaging and cell counting

Fluorescent imaging of the TAMRA–*f*PDCPD surface **26** was conducted on a Cytation 5 multichannel imaging platereader equipped with a 1.25x objective. A BioTek texas red filter cube with band widths of 576 – 600 nm for excitation and 610 – 685 nm for emission was used to measure fluorescence. The LED intensity on the instrument was set to 10, integration time was 1000 ms, and the gain setting was 24. The montage setting in the Cytation 5 software package was used to collect wide-field images. (Figure S4 and Table S2)

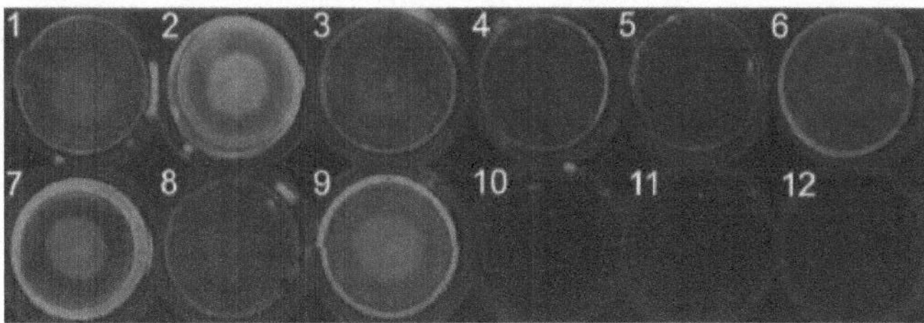

Figure S 4. Images of TAMRA functionalized surfaces with varying spin speeds, acceleration and polymer volume. Also shown are the methyl ester *f*PDCPD 4, partially hydrolyzed fPDCPD 24 and non-polymer-coated methacrylated glass. Images were collected using a Cytation 5 multichannel plate reader using the Texas Red Filter Cube and a 1.25x objective. The instrument's Gen 3.05 software was used, and an automated experiment was conducted to ensure identical imaging parameters for each surface. Numerical labels refer to the samples identified in Table S2.

Fluorescent imaging analysis and cell counting was accomplished using a Cytation 5 multichannel imaging plate reader. Automated agitation of multiwell plates was done using the plate shaker function on a SpectraMax M5 multichannel platereader. Mammalian cell counts were carried out using the standard Cytation 5 Imaging Reader software (v. 3.0.3), using a digital phase contrast of 30 µM, deconvolution of 1 SD, kernal radius of 20 px, cell radius parameters 11 µM minimum to 50 µM maximum, and a threshold setting of 6000.

Table S 2. Spin-coating parameters for the *f*PDCPD coated glass cover slides in Figure S4. Various volumes of 4wt% linear *f*PDCPD were spin-coated onto methacrylated glass slides with caring maximum speed and acceleration.

Image #	Spin Time/s	Max Speed/RPM	Acceleration/(RPM/s)	Volume Spin-Coated/µL
1	30	3000	750	30
2	30	1000	750	50
3	30	3000	1000	25
4	30	3000	500	50
5	30	3000	1000	30
6	30	2000	750	50
7	30	1000	500	50
8	30	2000	750	25
9	30	2000	1000	50
10 (Methyl Ester Negative Control)	30	3000	1000	50
11 (Carboxylic Acid Negative Control)	30	3000	1000	50

12 (Methacrylated Glass Negative Control)	--	--	--	--

7.3.2 Optical imaging analysis

Optical density readings for bacteria were conducted on a SpectraMax M5 platereader.

7.3.3 Quantification of accessible *f*DCPD–CO$_2$H groups in polymer 24

Samples of polymer **7** were prepared on 15 mm micro glass cover slides (surface area = 1.76 cm^2) as described above, using a variety of different concentrations of linear *f*PDCPD **2** (2 wt%, 3 wt% or 4 wt%) in 1:1 toluene:CH$_2$Cl$_2$, and crosslinking for two different times (16 hours or 48 hours) prior to hydrolysis (NaOH in 1:1 H$_2$O:MeOH, as described above).

The prepared samples were first incubated in a TBO solution (84 mg/mL; 6.4 µmol/mL) for 1 minute, then immediately washed (dipped with slight swirling) in a 10 wt% NaOH solution three times, and then air dried. Dried samples were then incubated in 3 mL of 50% acetic acid for 10 minutes with occasional stirring. The solution was then diluted 10x, and the absorbance was recorded at 630 nm, using a 50 wt% acetic acid solution as a blank. A non-saponified surface (i.e. **23**) was used as a control to determine non-specific TBO absorption on the surface of the films and was subtracted from the saponified surfaces' absorption reading.

A stock solution of toluidine blue oxide (TBO) in 50% acetic acid at a concentration of 1.03 µmol/mL was freshly prepared. Two-fold serial dilutions were carried out to prepare a series of standards, with concentrations from 0.103 µmol/mL to 0.0016 µmol/mL. The absorbance at 630 nm was measured to create a calibration curve (R^2=0.9997) as shown below.

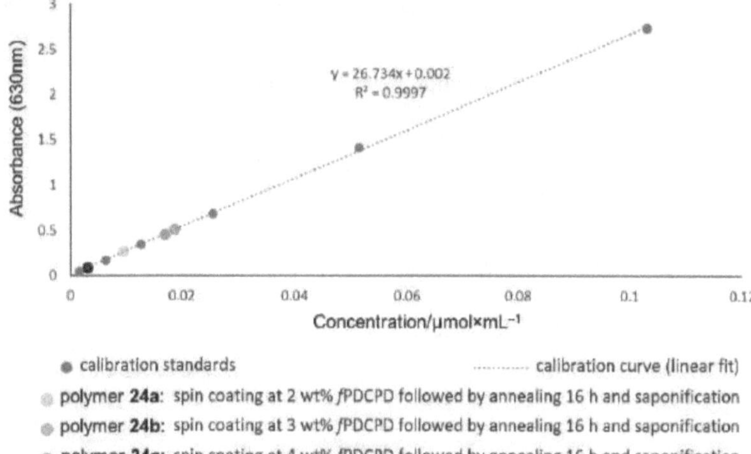

- calibration standards   ·············· calibration curve (linear fit)
- polymer **24a**: spin coating at 2 wt% *f*PDCPD followed by annealing 16 h and saponification
- polymer **24b**: spin coating at 3 wt% *f*PDCPD followed by annealing 16 h and saponification
- polymer **24c**: spin coating at 4 wt% *f*PDCPD followed by annealing 16 h and saponification
- polymer **24d**: spin coating at 4 wt% *f*PDCPD followed by annealing 48 h and saponification

Figure S 5. Calibration curve and quantification of *f*DCPD–COOH groups in polymer 24.

Table S 3. Accessible COOH Groups on the Surface of *f*PDCPD 24.

Sample	Control 23	*f*PCPD-CO₂H 24a	*f*PCPD-CO₂H 24b	*f*PCPD-CO₂H 24c	*f*PCPD-CO₂H 24d
	3 wt% loading; 16 h annealing	2 wt% loading; 1 6h annealing	3 wt% loading; 16 h annealing	4 wt% loading; 16 h annealing	4 wt% loading; 48 h annealing
µmol TBO per surface	0.126 ± 0.006	0.288 ± 0.049	0.511 ± 0.175	0.564 ± 0.221	0.097 ± 0.035
µmol TBO/cm²	0.071 ± 0.003	0.164 ± 0.028	0.290 ± 0.099	0.320 ± 0.125	0.055 ± 0.020
calculated* µmol COOH/cm²		0.092 ± 0.028	0.219 ± 0.100	0.249 ± 0.125	

*assuming 1:1 TBO:COOH binding

Table S3 summarizes the accessible concentration of COOH groups on the surface of each saponified sample. The data show that the amount of accessible carboxylate groups increases with increasing loading (i.e. a greater amount of *f*PDCPD **7** used in the spin-coating step), and decreases with crosslinking time. This is most probably due to a lack of swelling for the more-crosslinked samples. As we showed previously,[192] crosslinking density can be controlled by thermal annealing time. As annealing time (and therefore crosslinking density) is increased, the amount of solvent (and therefore TBO) that can penetrate into the polymer matrix is reduced. As a result, the amount of solvent-accessible groups naturally decreases.

7.3.4 HeLa cell adhesion measurements

HeLa cells were cultured with RPMI-1640 media supplemented with L-glutamine and 10% fetal bovine serum (FBS) at 37 °C. Adherent cells were removed from the surface of growth chambers (T-150 flasks) through trypsinization using 0.05 % trypsin–EDTA, counted using a hemocytometer, and diluted to afford a 5×10^4 cell/mL suspension. 1 mL of the prepared HeLa cell stock suspension was added to each well of a series of 24-well tissue culture plates, where each well either contained a premade cover slide affixed with one of the polymers described above, or contained a premade cover slide coated only with the 3-(trimethoxysilyl) propyl methacrylate reagent, or was empty (i.e. contained no cover slide). The plates were incubated for 48 hours at 37 °C and 5 % CO_2 to allow cells to adhere to the treated cover slides, after which each well was imaged using a Cytation 5 Imaging Reader. Adherent cell populations were quantified using the software packaged with the platereader. Media was then removed from all sample wells using suction filtration, and replaced with 1 mL per well of PBS buffer. The plate was shaken for 3 seconds using the plate shaker function on a SpectraMax M5 multichannel platereader. This two-step washing procedure was repeated twice more, after which 1 mL of fresh RPMI-1640 media containing 10 % fetal bovine serum was added to each well. The cell imaging protocol was repeated to determine the number of cells that remained attached to each cover slide throughout the three FBS washes. A ratio of the number of cells before and after the washing protocol was used to determine the adhesive properties of each surface.

7.3.5 *E. coli* growth measurements

Methacrylated glass cover slides, *f*PDCPD–chloramphenicol-coated cover slides, *f*PDCPD–CO_2Me-coated cover slides **23** and saponified *f*PDCPD-coated cover slides **24** were prepared as described above. These were inserted into the wells of 24-well tissue culture plates using an identical procedure to that described above for the HeLa cell experiments, once again using epoxy to ensure adhesion to the bottom of the plate. *E. coli* K12 was grown in baffled Erlenmeyer flasks containing lysogeny broth (LB) at 37 °C, under an atmosphere of 5 % CO_2 and constant 200 rpm agitation. 1 mL of *E. coli* suspension in LB media at an initial OD_{600} of 0.3 was added to each well of a series of 24-well tissue culture plates, where each well either contained a premade cover slide affixed with one of the polymers described above, or contained a premade cover slide coated only with the 3-(trimethoxysilyl) propyl methacrylate reagent, or was empty (i.e. contained no cover slide). As a positive control for growth inhibition, 1 mg/mL free chloramphenicol was added to an additional set of wells containing *E. coli* but no cover slides. The plates were maintained at 37 °C for the duration of the experiment. At regular time points (every 30 minutes), 200 µL of *E. coli* from each well of the 24-well plates was transferred to a fresh 96-well plate, and an optical density measurement was taken at 600 nm. A two-second shake was used prior to each absorbance measurement to ensure an even suspension of the *E. coli* cells.

7.4.0 General synthetic methods
7.4.1 Monomer synthesis

Synthesis of monomer 4

Monomer **4** was synthesized together with regioisomer **3** as previously described.[194] The polymerizable regioisomer (*f*DCPD) was isolated by reacting the mixture with 1,3-diaminopropane (5 eq.), trimethylamine (5 eq.), and DBU (0.1eq.) in dichloromethane for 16 hours, affording a selective conjugate addition to the unwanted regioisomer, the adduct of which was removed using an acidic aqueous extraction. Spectral data were consistent with previous reports.[194]

^{1}H NMR (300 MHz, CD$_2$Cl$_2$) δ 6.54-6.51 (m, 1H), 6.01 (dd, J = 5.7, 3.0 Hz, 1H), 5.91 (dd, J = 5.7, 3.0 Hz, 1H), 3.66 (s, 3H), 3.35-3.29 (m, 1H), 2.96-2.86 (m, 2H), 2.86-2.76 (m, 1H), 2.41 (ddt, J = 17.2, 10.6, 2.0 Hz, 1H), 1.89 (dtd, J = 17.2, 4.0, 2.0 Hz, 1H), 1.48 (dt, J = 8.0, 2.0 Hz, 1H), 1.28 (d, J = 8.2 Hz, 1H).

Large-scale synthesis of regioisomeric monomer mixture

Scheme S 1. Summarized method for producing *f*PDCPD at a big scale, based on selective polymerization from a chromatographically purified monomer mixture.

Freshly cracked cyclopentadiene (272.04 g, 4.12 mol), was slowly added to a solution of NaH (99 g, 4.12 mol) in dried tetrahydrofuran (THF, 2.06 L) at 0 °C over 1 hour. Dimethylcarbonate (1750 mL, 20.6 mol) was then added to the resulting 2M NaCP solution and the mixture was heated to 40 °C for 12 hours. Volatiles were removed in vacuo, and freshly cracked cyclopentadiene (544.08 g, 8.24 mmol), isopropanol (3 L) and H_2SO_4 (140 mL, 2.58 mol, 0.6 eq.) were added. The mixture was stirred for 48 hours. Volatiles were then removed in vacuo and the resulting black mixture was dissolved in hexanes and washed with deionized water. Volatiles were then removed in vacuo, affording an oil. This oil was then heated at 50 °C under vacuum (0.1 mmHg) for 4 hours to remove all remaining dicyclopentadiene, resulting in a black oil. Yield: 468 g, 60%; as a crude mixture of regioisomers **4**, **3** and **21**. Spectroscopic data for **4** and **9** were identical to that reported previously.[192-194]

Removal of isomer 3 through conjugate addition with 1,3-diaminopropane

40 g of the unpurified mixture of monomers described above (after removal of residual dicyclopentadiene at 50 °C and 0.1 mm Hg) was stirred in a round-bottom flask. The non-polymerizable regioisomer (~23 g, 0.12 mol) was reacted by adding triethylamine (92.4 mL, 5 eq.), DBU (9.8 mL, 0.5 eq.), and 1,3-diaminopropane (58.1 mL, 5 eq.) at 0 °C with stirring for 24 hours under argon. The resulting mixture was dissolved in diethyl ether and washed with 1M HCl, followed by saturated aqueous $NaHCO_3$ and saturated aqueous NaCl. Volatiles were removed in vacuo, resulting in the isolation of a dark brown oil. The oil was dried under high vacuum to remove solvent residue, affording 16 g (>90% recovery) of a crude mixture of regioisomers **4** and **31**.

Chromatographic purification of monomer 4

The crude mixture of **4** and **21** (following removal of **3** with 1,3-diaminopropane) was dissolved in hexanes and loaded onto a silica gel column. Elution with 20:1 hexanes:ethyl acetate followed by concentration in vacuo provided monomer **4** as a light yellow oil with an estimated purity (by NMR) of ≥90%. Spectroscopic data were identical to that reported previously.[192-194]

Synthesis of monomer 15

Sodium cyclopentadienide (2.0 M in THF, 16 mL, 0.032 mol) and dimethyl carbonate (14 mL, 0.16 mol) were combined, and the solution was stirred at 60 °C under a N_2 atmosphere for 16 hours. The reaction mixture was then cooled to room temperature, H_2O (50 mL) was added, and stirring was continued for 10 minutes. The mixture was then extracted with CH_2Cl_2. Volatiles were removed in vacuo, and the residue was redissolved in THF (20 mL). Cp-d_6 (5.1 g, 0.07 mol) was then added, and the solution was stirred overnight at 35 °C. The reaction mixture was then concentration in vacuo, D_2O (20 mL) was added, and the product was extracted with CH_2Cl_2, dried with Na_2SO_4, filtered, and volatiles were removed in vacuo to afford a yellow oil. The crude product was dissolved in hexanes and loaded onto a silica gel column. Elution with 9:1 hexanes:ethyl acetate provided the expected mixture of regioisomers. The isomers were separated by reacting the unwanted regioisomer with 1,3-diaminopropane (5 eq.), triethyl amine (5 eq.), and DBU (0.1 eq.) in CH_2Cl_2 (25 mL) at 60 °C overnight. The solution was washed with 1 M DCl/D_2O, and the organic layer was dried with Na_2SO_4, filtered, and concentrated in vacuo. The resulting oil was then loaded onto a silica column and eluted with toluene followed by ethyl acetate. Overall yield: 104 mg, 1.7%. Compound identity was confirmed by comparison of ^1H and ^2H NMR spectra to the ^1H spectra of the known, undeuterated monomer,**4** (see Figure A2-4 in Appendix).

^1H NMR (300 MHz, $CDCl_3$) δ 6.53-6.52 (m, 1H), 3.66 (s, 3H), 3.33-3.30 (m, 1H), 2.83-2.78 (m, 1H), 2.44-2.37 (m, 1H), 1.91-1.86 (m, 1H) ; ^2H NMR (500 MHz, CH_2Cl_2) δ 6.00 (s, 1D), 5.90 (s, 1D), 2.86 (s, 1D), 2.83 (s, 1D), 1.39 (s, 1D), 1.21 (s, 1D); $^{13}C\{^1H\}$ NMR (125 MHz, $CDCl_3$) δ 165.8, 144.6, 137.2, 55.0, 51.4, 41.2, 33.7 (deuterium labeled carbons were very weak in the ^{13}C spectrum).

Synthesis of monomer 17

Cp-d_6 (2.610 g, 0.036 mol) was slowly added to a suspension of NaH (0.957 g, 0.040 mol) in THF (50 mL) at 0 °C. The reaction mixture was then warmed to room temperature, and dimethyl carbonate (16.2 g, 0.18 mol) was added. The reaction mixture was stirred for 16 hours at 50 °C under a N_2 atmosphere. The mixture was then cooled to room temperature, and D_2O (20 mL) was added. Stirring was continued for 10 minutes, after which the mixture was extracted with CH_2Cl_2, and volatiles were removed in vacuo. THF (100 mL) and freshly cracked cyclopentadiene (4.752 g, 0.072 mol) was then added, and the mixture was stirred overnight at 30 °C. The reaction mixture was then concentration in vacuo and loaded onto a silica gel column. Elution with 9:1 hexanes:ethyl acetate provided the expected mixture of regioisomers. The isomers were separated by reacting the unwanted regioisomer with 1,3-diaminopropane (5 eq.), triethyl amine (5 eq.), and DBU (0.1 eq.) in THF (50 mL) at 50 °C overnight. Volatiles were removed in vacuo. The residue was redissolved in Et_2O and washed with 1M DCl/D_2O, and the organic layer was then dried with Na_2SO_4, filtered, and concentrated in vacuo. The resulting oil was loaded onto a silica gel column and eluted with toluene. Overall yield: 146 mg, 2.1 %. Compound identity was confirmed by comparison of ^1H and ^2H NMR spectra to the ^1H spectra of the known, undeuterated monomer, **4** (see Figure A5-9 in Appendix A).

^1H NMR (300 MHz, $CDCl_3$) δ 6.04-6.01 (m, 1H), 5.94-5.91 (m, 1H), 3.67 (s, 3H), 2.93 (br s, 1H), 2.89 (br s, 1H), 1.49 (dt, J = 8.2, 1.8 Hz, 1H), 1.29 (d, J = 8.3 Hz, 1H); ^2H NMR (500 MHz, CH_2Cl_2) δ 6.46 (s, 1D), 3.24 (s, 1D), 2.71 (s, 1D), 2.28 (s, 1H), 1.76 (s, 1D); $^{13}C\{^1H\}$ NMR (125 MHz, $CDCl_3$) δ 165.6, 135.6, 133.5, 132.9, 51.6, 51.2, 50.3, 46.2, 45.4 (deuterium labeled carbons were very weak in the ^{13}C spectrum); HRMS (ESI): $[M+H]^+$ $C_{12}H_{10}D_5O_2$ Calcd 196.1380; Found 196.1382.

Synthesis of monomer 19

Compound **4** (0.416 g, 2.18 mmol), 1,3-di-tert-butyl-2-(neopentyloxy)-2,3-dihydro-1H-1,3,2-diazaphosphole (70 mg, 0.024 mol), pinacolborane (3.27 mmol), and CH_3CN (5 mL) were stirred at room temperature for 3 days. The reaction was exposed to the atmosphere and volatiles were then removed in vacuo. The resulting oil was loaded onto a silica gel column and eluted with toluene to afford the desired product. Yield: 223 mg, 53%.

^{1}H NMR (300 MHz, CDCl$_3$) δ 6.18 (t, J = 2.0 Hz, 2H), 3.62 (s, 3H), 2.78-2.74 (m, 3H), 2.67-2.61 (m, 2H), 1.93-1.84 (m, 2H), 1.68 (dt, J = 8.0, 1.8 Hz, 1H), 1.52 (d, J = 8.0 Hz), 1.19-1.08 (m, 2H) ; ^{13}C{^{1}H} NMR (125 MHz, CDCl$_3$) δ 175.4, 136.7, 54.2, 51.3, 48.8, 46.6, 45.3, 32.5

Synthesis of monomer **29, 30, 31** and **32**

Scheme S 2. Synthetic route of producing compound 29 and 30 from 4.

116 mg of purified ƒDCPD–CO2Me monomer **4** (1; 0.61 mmol) was added to 1.6 mL of a 7 wt% solution of sodium hydroxide (NaOH) in 1:1 MeOH : distilled water. The solution was heated to 50 °C and left to react overnight. After cooling to room temperature, 1 M HCl was used to precipitate the desired carboxylic acid intermediate. After partial concentration in vacuo to remove methanol, the product was extracted with ethyl acetate to afford 102 mg (0.58 mmol; 95 %) of crude ƒDCPD–CO$_2$H (compound **29**) as a white powder. The acid intermediate was then immediately dissolved in 1.5 mL of dichloromethane, to which 0.42 mL of thionyl chloride (5.8 mmol; 10 eq.) was added under argon. The solution was left to react overnight at room temperature with stirring, after which the solvent was removed in vacuo to afford 100 mg (0.51 mmol; 89 %) of crude ƒDCPD–COCl (compound **30**) as a clear, yellow-brown oil.

Scheme S 3. Synthetic route of producing compound 31.

Crude ƒDCPD–COCl (compound **30**) (195 mg; 1.00 mmol) was dissolved in 2 mL of CH$_2$Cl$_2$. N,N-Dimethylaminopyridine (12.3 mg; 0.1 mml; 0.1 eq.) was added and the mixture was cooled to 0 °C. 1-Octanol (1.6 mL; 10 mmol; 10 eq.) and triethylamine (1.4 mL; 10 mmol; 10 eq.) was added to the stirring solution. After 5 min, the ice bath was removed, and the reaction mixture was allowed to warm to room temperature over 12 h. The reaction mixture was washed with 1 M HCl and the aqueous layer was back-extracted twice with CH$_2$Cl$_2$. The combined organic layers were dried over MgSO$_4$ and then evaporated to afford the crude ƒDCPD–octyl ester product (compound **32**). This was subjected to flash-column chromatography over silica gel, using 2.5:1 hexanes:CH_2Cl_2 as the eluent. The desired product was obtained as a clear yellow oil in a yield of 50%.

^{1}H NMR (300MHz, CDCl$_3$) δ 6.54-6.52 (m, 1H), 6.03 (dd, J = 5.6, 2.9 Hz, 1H), 5.94 (dd, J = 5.6, 2.9 Hz, 1H), 4.06 (t, J = 6.6 Hz, 2H), 3.36-3.30 (m, 1H), 2.94-2,90 (m, 2H), 2.86-2.80 (m, 1H), 2.42 (ddt, J = 17.2, 10.3, 1.9 Hz, 1H), 1.94-1.86 (m, 1H), 1.62 (t, J = 6.6 Hz, 2H), 1.50 (dt, J = 8.2, 1.8 Hz, 1H), 1.43-1.14 (m, 11H), 0.88 (t, J = 6.6 Hz, 3H); ^{13}C NMR (75 MHz, CDCl$_3$) δ 165.4, 144.1, 137.5, 135.7, 133.0, 64.2, 55.0, 50.3, 46.4, 45.6, 41.3, 33.6, 31.8, 29.7, 29.2, 28.6, 26.0, 22.6, 14.1; IR (cm^{-1}, film) 3065, 2959, 2924, 2854, 1716, 1638, 1461, 1383, 1343, 1280, 1232, 1132, 1095, 1017, 917, 755, 733, 707, 518, 470; HRMS: [M+H]$^+$ calculated for $C_{19}H_{29}O_2$ 289.2162; found 289.2161.

Scheme S 4. Synthetic route for producing compound 32.

Crude ƒDCPD–COCl (compound **30)** (195 mg; 1.00 mmol) was dissolved in 2 mL of CH$_2$Cl$_2$. N,N-Dimethylaminopyridine (12.3 mg; 0.1 mml; 0.1 eq.) was added and the mixture was cooled to 0 °C. Tetraethylene glycol (1.7 mL; 10 mmol; 10 eq.) and triethylamine (1.4 mL; 10 mmol; 10 eq.) was added to the stirring solution. After 5 min, the ice bath was removed, and the reaction mixture was allowed to warm to room temperature overnight. A crude NMR spectrum was collected to confirm that the reaction had gone to completion, after which the reaction mixture was washed with 1M HCl and saturated aqueous sodium bicarbonate. Following each wash, the aqueous layer was extracted twice with CH$_2$Cl$_2$. The combined organic layers were dried over MgSO$_4$ and then evaporated to afford the crude fDCPD–TEG ester product. This was subjected to flash-column chromatography over silica gel, using 100 % CH$_2$Cl$_2$ followed by 2:1 CH$_2$Cl$_2$:EtOAc and finally 30:1 CH$_2$Cl$_2$:MeOH as the eluent. The desired product was obtained as a clear, colorless oil.

7.4.2 General polymer synthesis

Linear polymer 7 (precursor to crosslinked polymer 14)

Polymer **7** and the linear polymer precursors to **27** and **28** were prepared by adding 1 mol% Grubbs 2nd generation catalyst (G2) to 50 mg of the monomer **4** in 1.0 g dichloromethane in a N$_2$ filled glovebox. Polymerization was allowed to proceed for 1 hour with stirring, after which NMR analysis indicated complete consumption of monomer. The reaction was quenched by the addition of 0.5 mL ethyl vinyl ether, and the polymer product was precipitated by the addition of hexanes. Removal of solvent *via* centrifugation (3750 rpm at 4 °C) followed by drying in vacuo afforded the target linear polymers. This purification was repeated twice for each sample. Volatiles were removed in vacuo to afford a yellow-brown solid. Isolated yields were consistently between 70–80 %.

^{1}H NMR (300 MHz, CDCl$_3$) δ 6.59-6.55 (br, 1H), 5.51-5.42 (br, 2H), 3.70 (s, 3H), 3.40 (br, 1H), 2.96 (br, 2H), 2.68 (m, 2H), 2.50 (br, 2H), 1.67 (br, 1H), 1.28 (br, 1H); ATR-FTIR (cm^{-1}): 2930, 2854, 1713, 1635, 1435, 1356, 1265, 1198, 1093, 973, 917, 733; Raman (cm^{-1}): 3063, 3001, 2949, 2855, 1714, 1636, 1443, 1303, 1263, 1199, 1033, 1058, 971, 856.

Reaction injection molding of homopolymers and copolymers 39-42

Functionalized dicyclopentadiene compound **4** (purified as indicated below) and or non-functionalized dicyclopentaidne compound **6** were combined under inert atmosphere in a glove box with the G2 catalyst (1 mol %) in a glass vial. The mixture was stirred with a spatula for 30 seconds, then taken up in a plastic syringe, and immediately transferred through sprue holes (~3 mm) cut into a ¼" aluminum top plate, into an aluminum mold with a depth of ¼". Test samples prepared in this way had a height of approximately 1" and a variable width and shape, as indicated in Figure 19. The samples were allowed to polymerize for 24 hours at room temperature in the glove box, after which the mold (still containing the polymer samples) was removed from the glove box and placed in an oven at 135 °C to effect thermal curing. Different curing times (30 minutes to 144 hours) provided samples with different degrees of coloring. The mold was then disassembled by removing the top and bottom plates, and the samples were removed. To facilitate removal of the polymer samples, the aluminum mold (as well as the associated top and bottom plates) was treated with a Teflon-containing spray prior to use.

Hardness and water contact angle measurement samples of homopolymers and copolymers 39-42

Samples were prepared following identical procedures to those described for the DMA samples, but using natural rubber RIM molds instead of aluminum RIM molds. The prepared samples were 14.5 mm in height and 14.5 mm in diameter.

FTIR, Raman, solid-state NMR, DSC, and TGA samples of copolymers 39-42

Dicyclopentadiene (compound **6**) was warmed by low-temperature heat gun for 10 minutes to maintain it in a liquid phase. Purified monomer (compound **4**) (1 mmol) was mixed with the warm dicyclopentadiene (1 mmol) in dichloromethane (DCM, 0.1 M) and stirred. G2 (1 mol%) was added to the stirring solution. Polymerization was allowed to proceed for 1 hour with stirring, after which polymers were precipitated by adding hexanes to dichloromethane solution. The product was isolated by centrifugation. The precipitation was repeated three times for each polymer sample to remove any impurities. Volatiles were removed in vacuo to afford a yellow or brown solid.

Polymer 16

^{1}H NMR (300 MHz, CDCl$_3$) δ 6.65-6.55 (br, 1H), 3.76 (s, 3H), 3.38 (br, 1H), 2.95 (br, 1H), 2.67-2.53 (m, 2H); ATR-FTIR (cm^{-1}): 3447, 2951, 2925, 2854, 2211, 2113, 1714, 1635, 1436, 1353, 1267, 1205, 1091, 735; Raman (cm^{-1}): 3061, 2947, 2856, 2213, 2120, 1721, 1631, 1446, 1287, 1106, 1056, 865, 782.

Polymer 18

^{1}H NMR (300 MHz, CDCl$_3$) δ 5.61-5.43 (m, 2H), 3.74 (s, 3H), 3.03 (br, 1H), 2.71 (br, 1H), 1.71 (br, 1H), 1.33 (br, 1H); ATR-FTIR (cm^{-1}): 2948, 2917, 2853, 2195, 2113, 1713, 1615, 1435, 1330, 1290, 1194, 1176, 1072, 975, 920, 761, 660; Raman (cm^{-1}): 3006, 2953, 2905, 2852, 2296, 2202, 2161, 2118, 1718, 1657, 1621, 1450, 1303, 1240, 1125, 1045, 831.

Polymer 20

^{1}H NMR (300 MHz, CDCl$_3$) δ 5.34 (m, 2H), 3.66-3.59 (m, 3H), 2.64-2.37 (m, 5H), 1.77-1.13 (m, 7H); ATR-FTIR (cm^{-1}): 3448, 2949, 2866, 1939, 1727, 1633, 1434, 1370, 1265, 1165, 1030, 972, 923, 748.

Preparation and methacrylation of glass cover slides

Matsunami 15 mm micro glass cover slides (chosen to fit precisely into the wells of a standard 24-well plate) were sonicated in a solution of Sparkleen detergent and deionized H$_2$O for one hour with occasional stirring. The slides were then thoroughly rinsed with deionized H$_2$O, followed by 95 % ethanol and then diethyl ether. Following evaporation of residual solvent at room temperature, the slides were then dried in an oven set to 110 °C. The glass slides were then evenly spread in a glass petri dish and incubated in a mixture of ethanol, acetic acid and 3-(trimethoxysilyl) propyl methacrylate for three minutes. The incubation medium was prepared by first diluting 1 part glacial acetic acid in 9 parts deionized water, then diluting 3 mL of this dilute acid solution with 96.5 mL of 95 % ethanol. To the resulting AcOH / EtOH solution was added 0.5 mL of the methacrylate reagent. After incubation, the slides were then thoroughly rinsed with 95 % ethanol and dried in vacuo.

fPDCPD spin coating and thermal crosslinking to afford 7, 23 or 24

A 1:1 solution of toluene:CH$_2$Cl$_2$ containing 4 wt% of linear polymer **7** (or for the experiments in Figure 21, the linear precursors to **23** or **24**) was prepared at room temperature in a sintered vial, using sonication to fully dissolve the polymer. 50 μL of the solution was then spin coated onto the methacrylated 15 mm round cover slides using a maximum speed of 3000 rpm, acceleration of 1000 rpm/s and a total application time of 30 seconds. The freshly spin-coated slides were then placed on glass petri dishes and crosslinked at 135 °C for 48 hours prior to use. For optimization details of the spin-coating protocol, see Table S2 above.

*Saponification of ƒPDCPD surfaces to afford **24***

Freshly crosslinked methyl ester-functionalized ƒPDCPD glass cover slides **23** were evenly placed on a glass petri dish and submerged in a 7 wt% solution of NaOH in 1:1 H_2O:MeOH for 30 minutes. The surfaces were then washed with water, followed by methanol, and finally dried in vacuo. FTIR analysis confirmed partial hydrolysis of the surface-accessible ester groups (Figure B9- in Appendix B)

*Formation of acyl chloride ƒPDCPD surfaces **25***

Freshly saponified surfaces **24** were separated into sintered vials sealed with inverted septa and parafilm. Under an atmosphere of argon, 1 mL of a solution containing 80 % CH_2Cl_2, 10 % DMF, and 10 % $SOCl_2$ was injected into each vial. The reaction mixture was left at room temperature for 2 hours, after which the surfaces were washed with CH_2Cl_2 and dried in vacuo.

*Synthesis of TAMRA-functionalized ƒPDCPD **26***

Freshly prepared acyl chloride surfaces **25** were separated into sintered vials sealed with inverted septa and parafilm. Under an atmosphere of argon, surfaces were submerged in 1 mL of a solution containing 0.15 mg/mL of 5-TAMRA-PEO3-amine (from 110 µL of a 1.43 mg/mL DMSO stock solution), 490 µL DMSO and 400 µL triethylamine. After 48 hours at room temperature, the slides were removed from the vials, washed three times with DMSO and placed into a 24-well plate for fluorescent imaging. FTIR analysis confirmed the formation of the target amide Figure B10 in Appendix B.

*Synthesis of chloramphenicol-functionalized ƒPDCPD **36***

Freshly prepared acyl chloride surfaces **25** were separated into sintered vials sealed with inverted septa and parafilm. Under an atmosphere of argon, surfaces were submerged in 1 mL of a solution containing 400 µL of triethylamine, 600 µL of DMSO, and 10 mg of chloramphenicol. After 48 hours at room temperature, the slides were removed from the vials, washed with acetonitrile followed by CH_2Cl_2, and dried in vacuo. FTIR analysis confirmed surface functionalization Figure B11 in Appendix B.

Formation of surfaces for HeLa cell adhesion assays

Methacrylated glass cover slides, ƒPDCPD–CO_2Me-coated cover slides **23** and saponified ƒPDCPD-coated cover slides **24** were prepared as described above. ƒPDCPD–perfluorooctyl ester-coated cover slides, as well as ƒPDCPD–RGD- and ƒPDCPD–RAD-coated cover slides, were prepared through an analogous procedure to that used for **7** (i.e. acyl chloride surfaces were exposed to either perfluorooctanol or else RGD or RAD amines, in the presence of triethylamine). All surfaces were disinfected with 70 % ethanol prior to being brought into a biological safety cabinet, and then were washed with phosphate-buffered saline (PBS) and air-dried for 20 minutes. Clear LePage Speed Set Epoxy was then applied to the bottom of each well of sterile 24-well tissue culture plates.

The prepared glass cover slides were laid over top of the epoxy and light pressure was applied to ensure strong adhesion. The epoxy was left to cure for 30 minutes prior to addition of cells.

Appendix

Appendix A: NMR Spectral Data for Selected Compounds

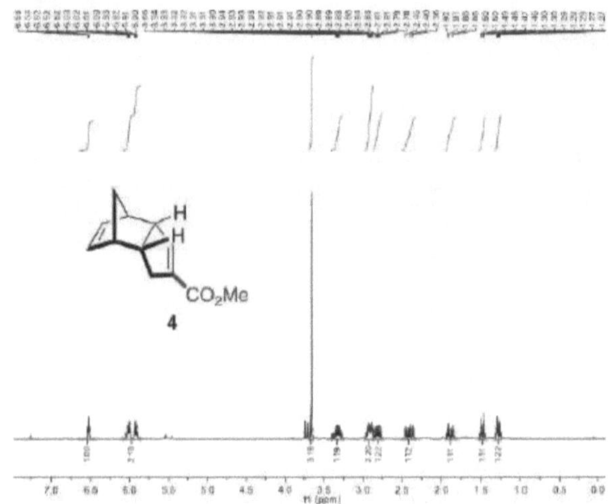

Figure A 1. ^{1}H NMR spectrum of monomer 4 (300 MHz, CDCl$_3$).

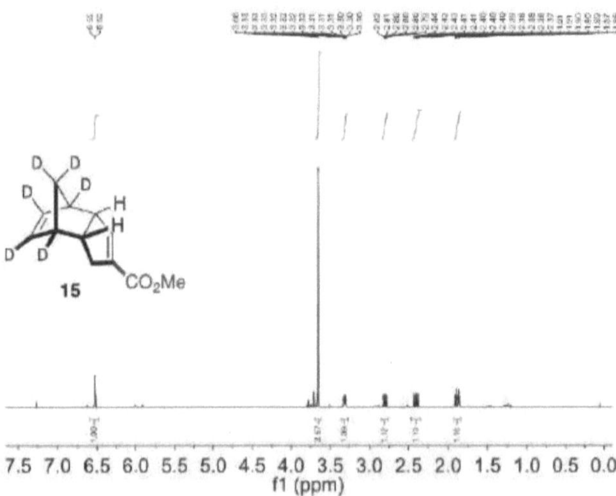

Figure A 2. ^{1}H NMR spectrum of monomer 15 (300 MHz, CDCl$_3$).

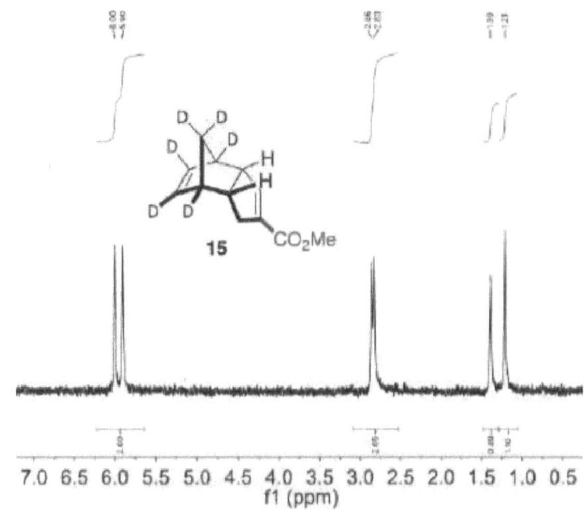

Figure A 3. ^{2}H NMR spectrum of monomer 15 (500 MHz, CH$_2$Cl$_2$).

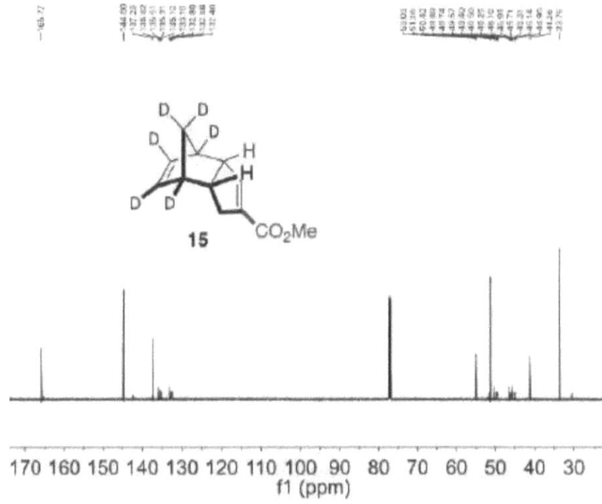

Figure A 4. ^{13}C{^{1}H} NMR spectrum of monomer 15 (125 MHz, CDCl$_3$).

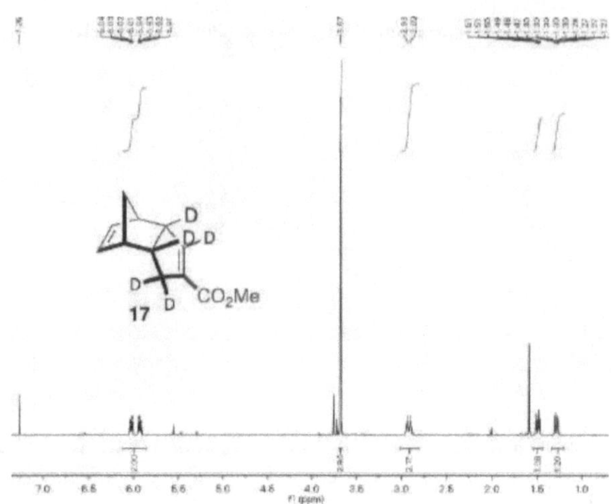

Figure A 5. ^{1}H NMR spectrum of monomer **17** (300 MHz, CDCl$_3$).

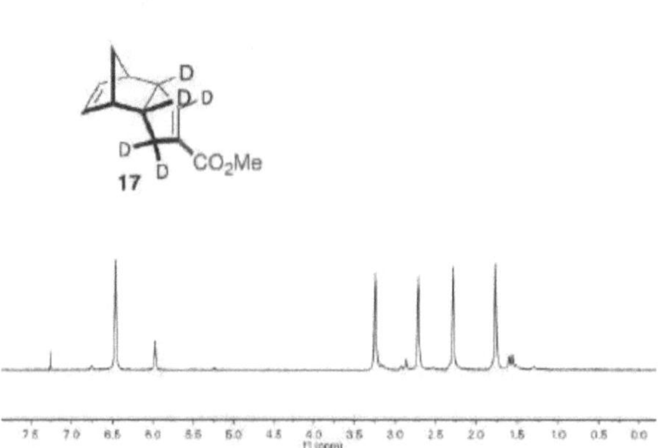

Figure A 6. ^{2}H NMR spectrum of monomer **17** (500 MHz, CH$_2$Cl$_2$).

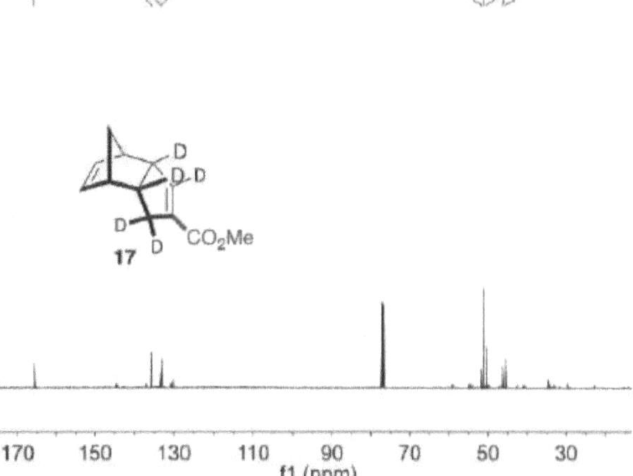

Figure A 7. ^{13}C{^{1}H} NMR spectrum of monomer 17 (125 MHz, CDCl$_3$).

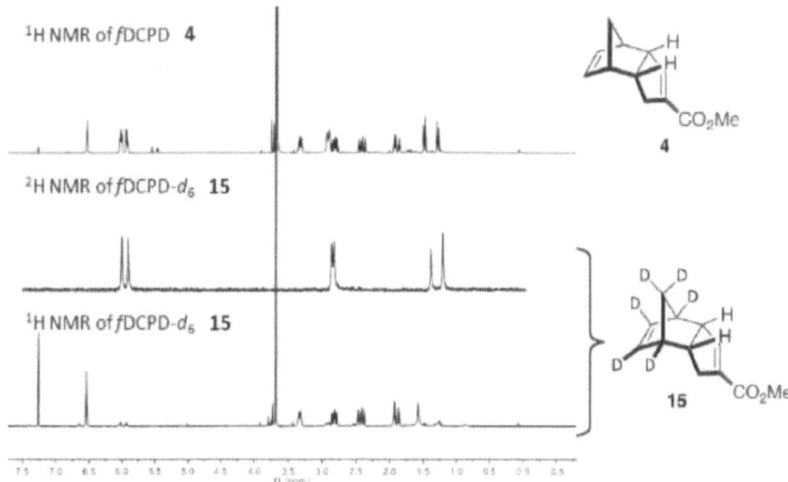

Figure A 8. Comparison of ^{1}H and ^{2}H NMR spectra for monomer 15 to the spectrum of the known, undeuterated monomer 4.

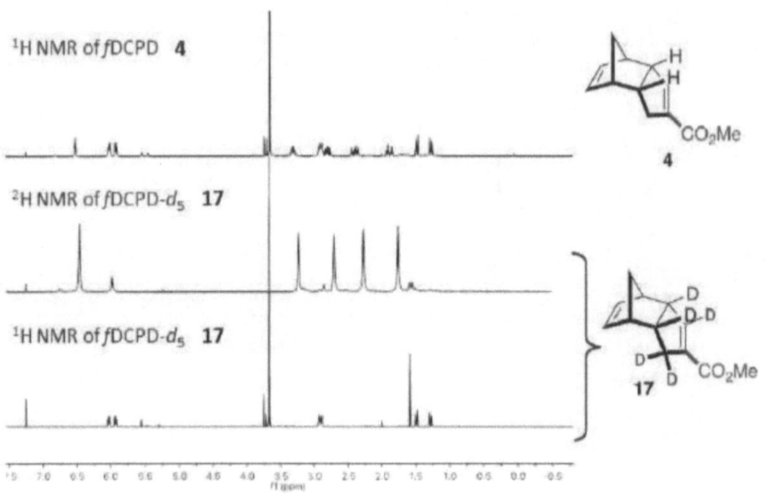

Figure A 9. Comparison of ^{1}H and ^{2}H NMR spectra for monomer 17 to the spectrum of the known, undeuterated monomer 4.

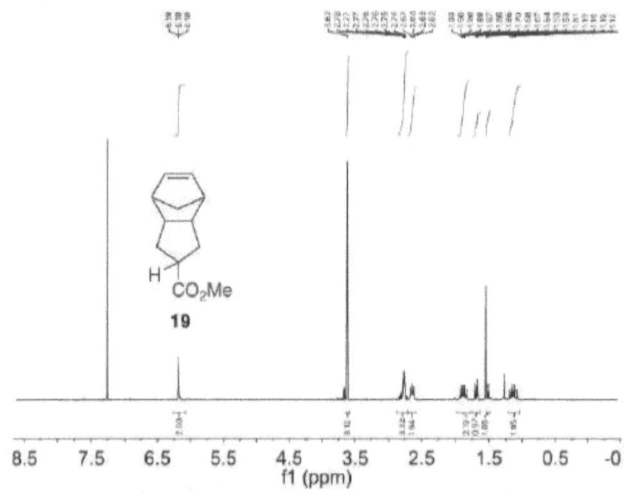

Figure A 10. ^{1}H NMR spectrum of monomer 19 (300 MHz, CDCl$_3$).

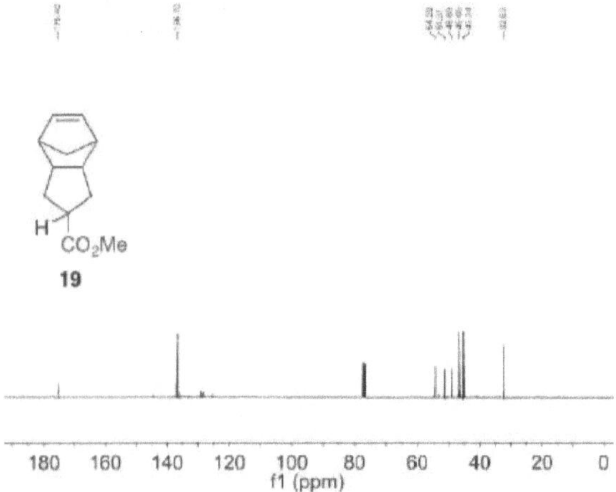

Figure A 11. $^{13}C\{^1H\}$ NMR spectrum of monomer 19 (125 MHz, CDCl$_3$).

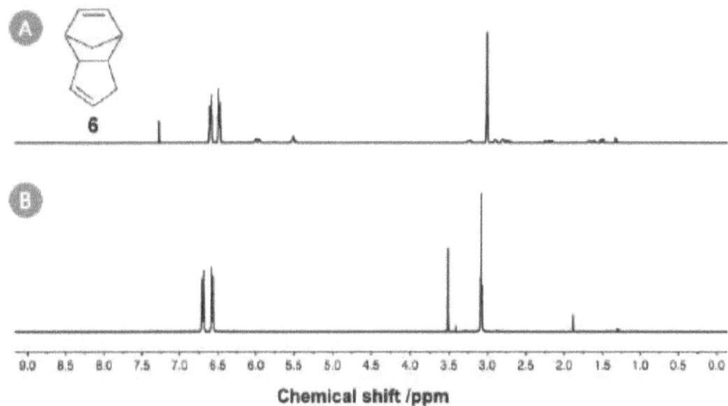

Figure A 12. Comparison of ^1H NMR spectrum for dicyclopentadiene (compound 6) recovered by distillation A to a spectrum for commercial dicyclopentadiene B.

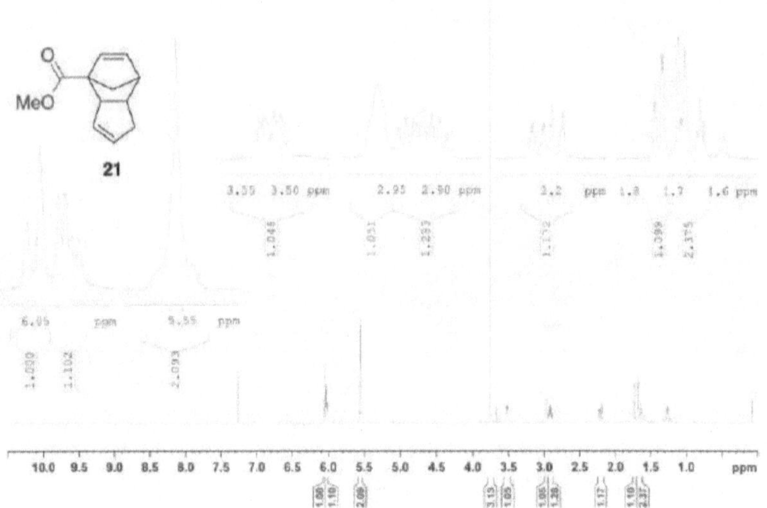

Figure A 13. ^{1}H NMR spectrum for compound 21 in CDCl$_3$, recorded at 500 MHz.

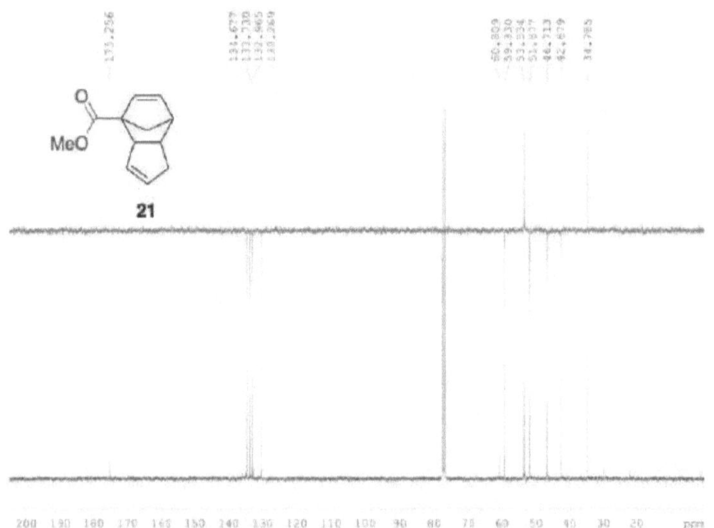

Figure A 14. ^{13}C NMR and DEPT-135 NMR spectra (in black and red, respectively) for compound 21 in CDCl$_3$, recorded at 75 MHz.

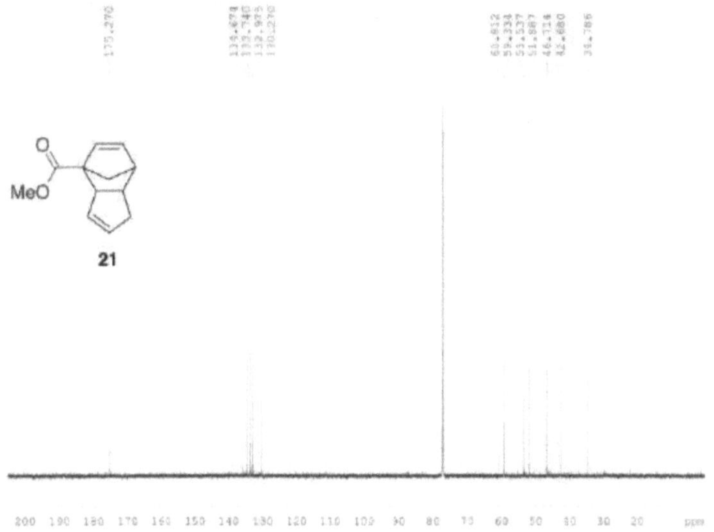

Figure A 15. ^{13}C NMR NMR spectrum for compound **21** in CDCl$_3$, recorded at 125 MHz.

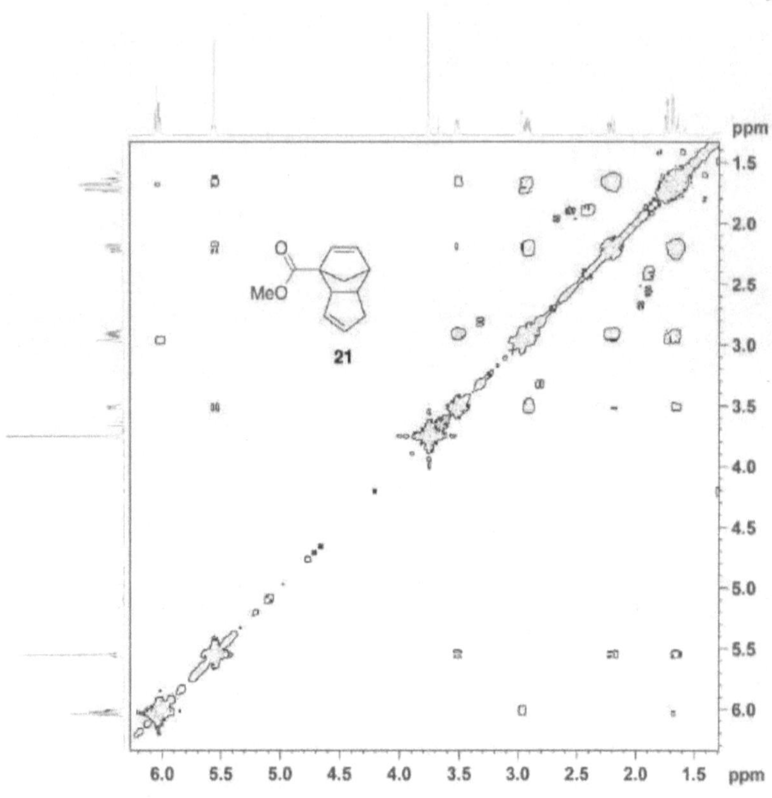

Figure A 16. COSY NMR spectrum (full range) for compound 21 in $CDCl_3$, recorded at 500 MHz.

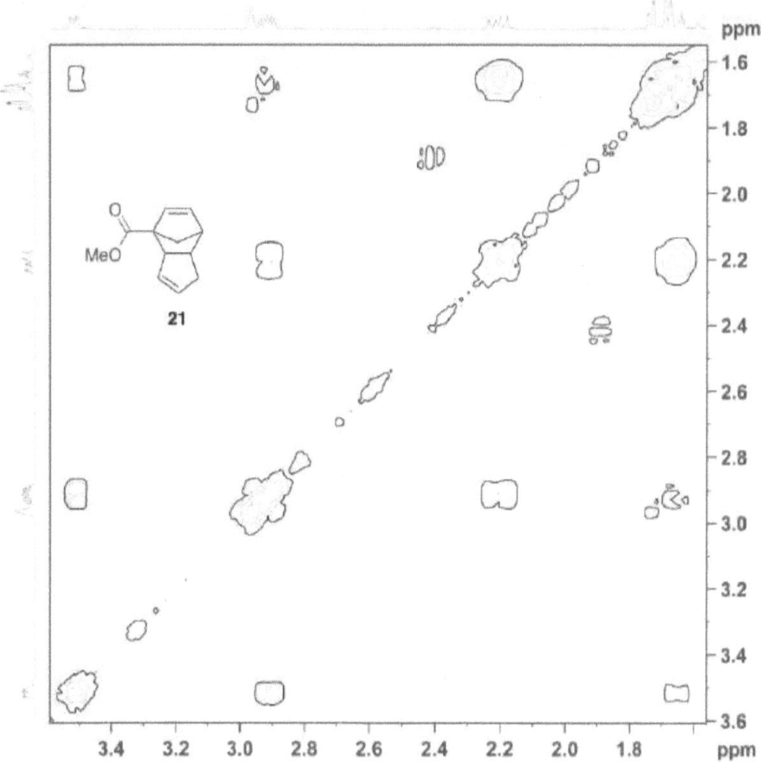

Figure A 17. COSY NMR spectrum (alkyl region) for compound 21 in CDCl₃, recorded at 500 MHz.

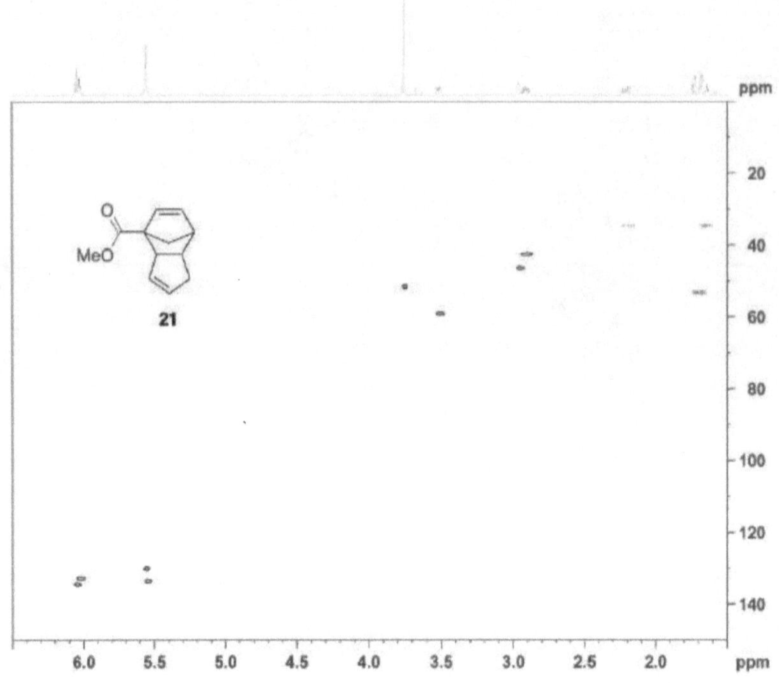

Figure A 18. HSQC NMR spectrum for compound 21 in $CDCl_3$, recorded at 500 MHz x 125 MHz.

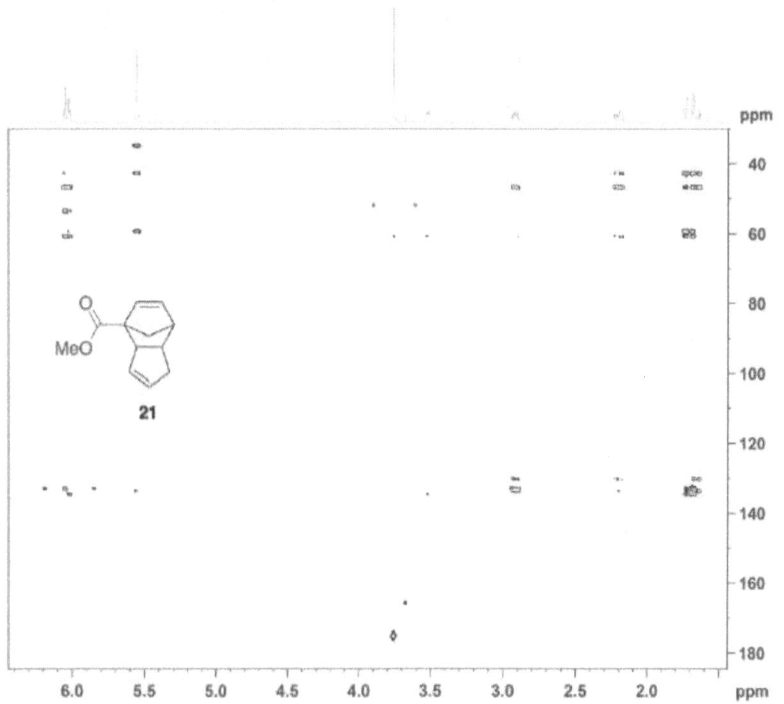

Figure A 19. HMBC NMR spectrum for compound 21 in CDCl$_3$, recorded at 500 MHz x 125 MHz.

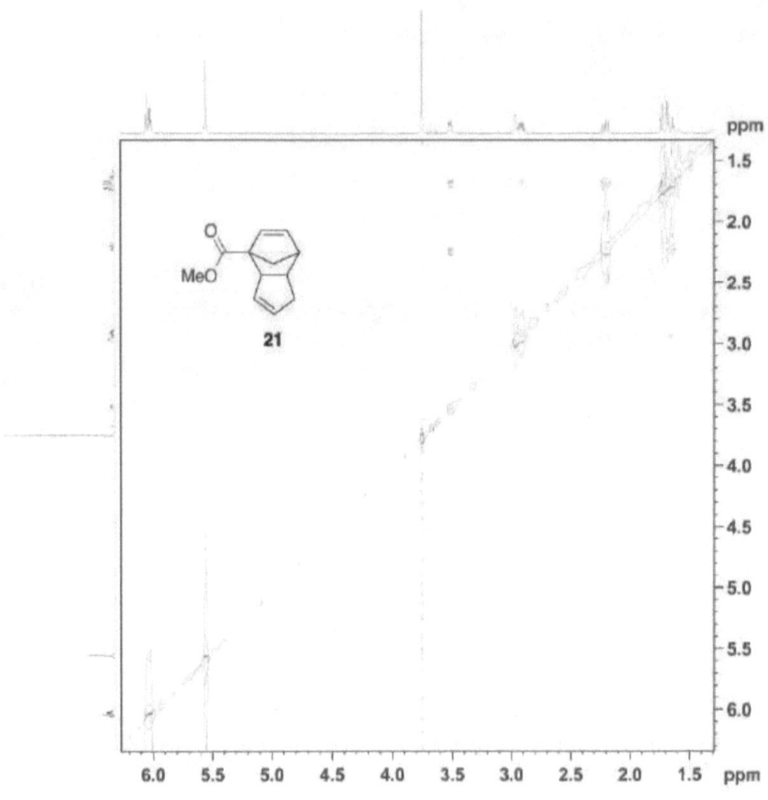

Figure A 20.NOESY NMR spectrum for compound 21 in CDCl$_3$, recorded at 500 MHz.

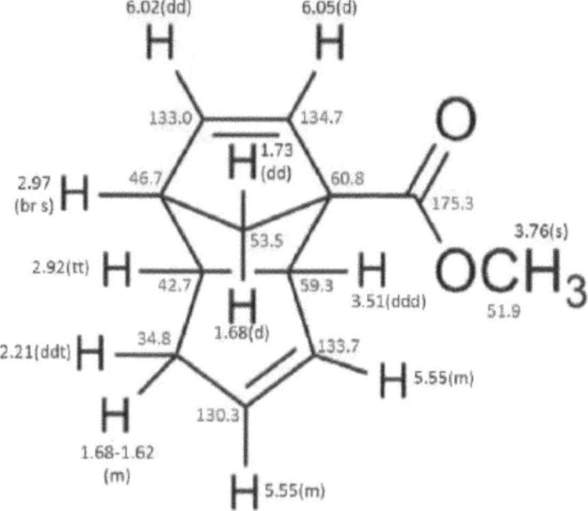

Figure A 21. NMR assignments for compound 21. Blue: ^{1}H NMR; red: ^{13}C NMR.

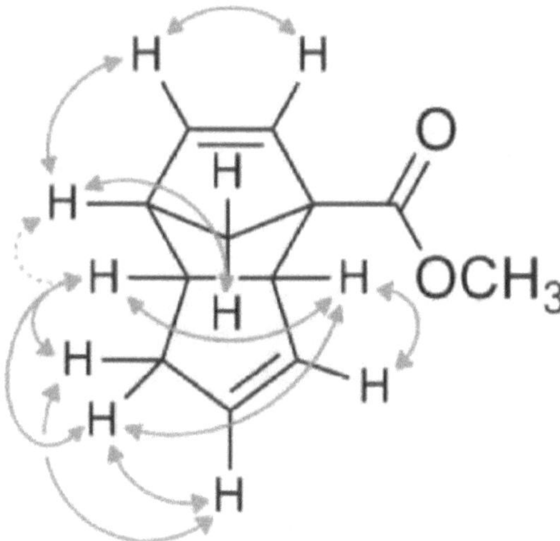

Figure A 22. Significant COSY correlations for compound 21, used to assign the structure.

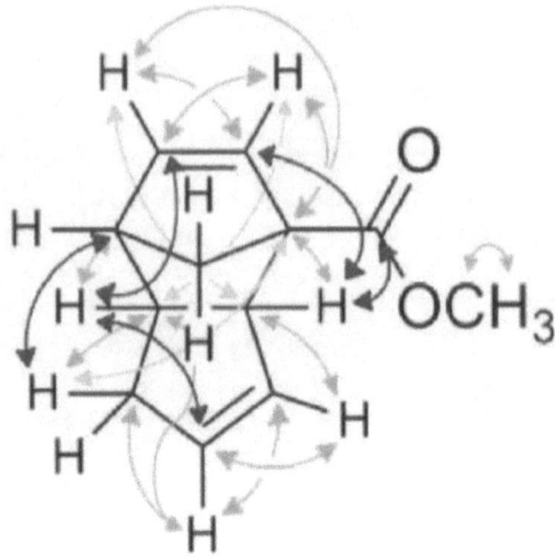

Figure A 23. Significant HMBC correlations for compound 21, used to assign the structure. Pink: confirmatory correlations establishing connectivity in the northern and southern halves of the molecule. Blue: most significant correlations establishing the relative orientation of the northern and southern fragments to one another. Grey: ^4J couplings present due to the constrained nature of the molecule.

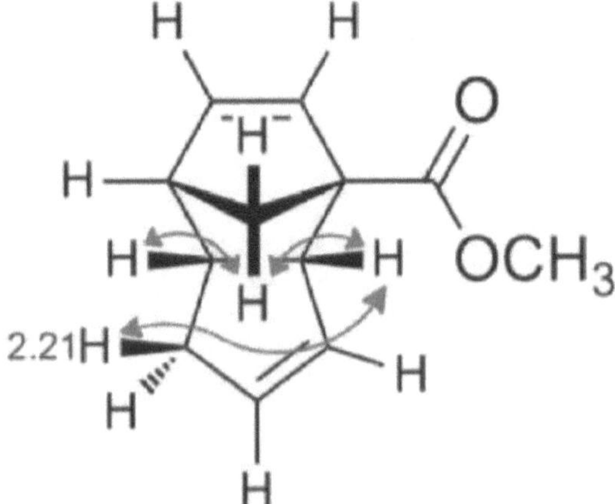

Figure A 24. Significant NOE correlations for compound 21, used to assign the structure.

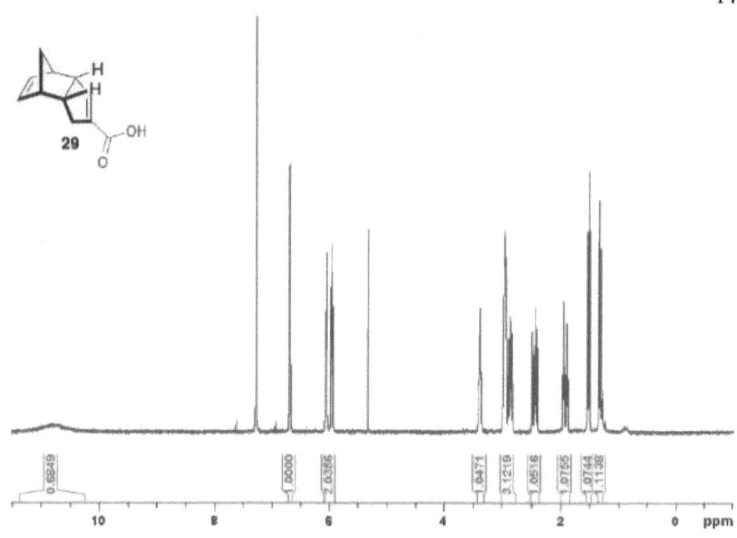

Figure A 25. ¹H spectrum for crude *f*DCPD–COOH in CDCl₃, recorded at 300 MHz.

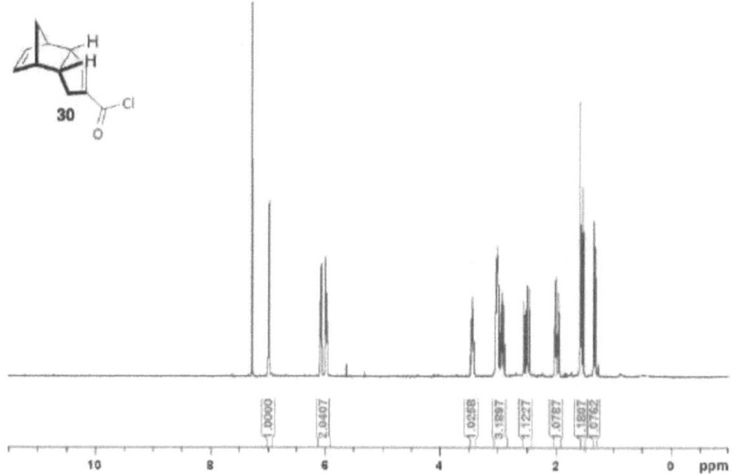

Figure A 26. ¹H NMR spectrum for crude *f*DCPD–COCl in CDCl₃, recorded at 300 MHz.

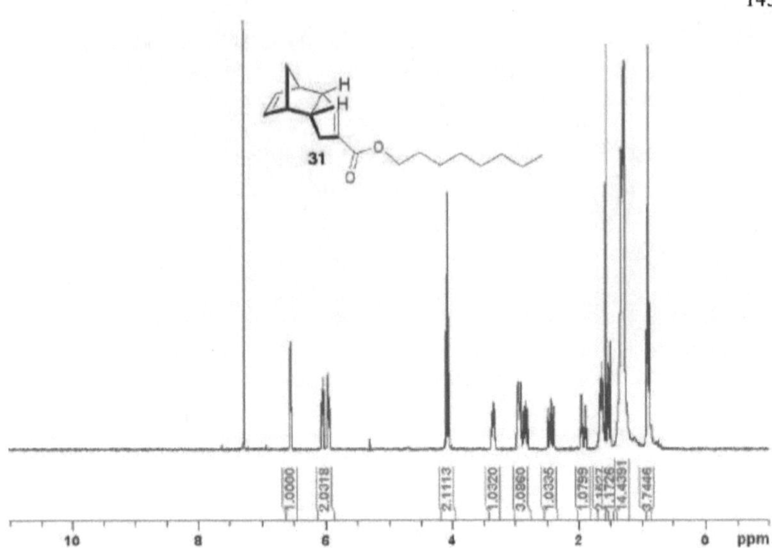

Figure A 27. ^{13}C NMR spectrum for *f*DCPD–octyl ester in CDCl$_3$, recorded at 300 MHz.

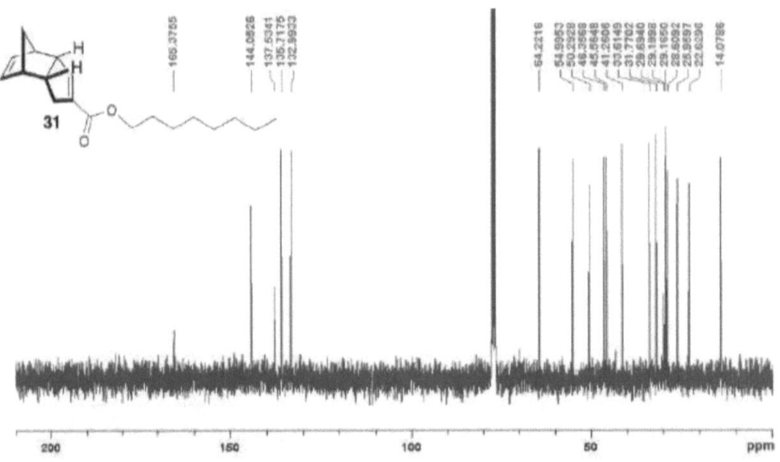

Figure A 28. ^{13}C NMR spectrum for *f*DCPD–octyl ester in CDCl$_3$, recorded at 75 MHz.

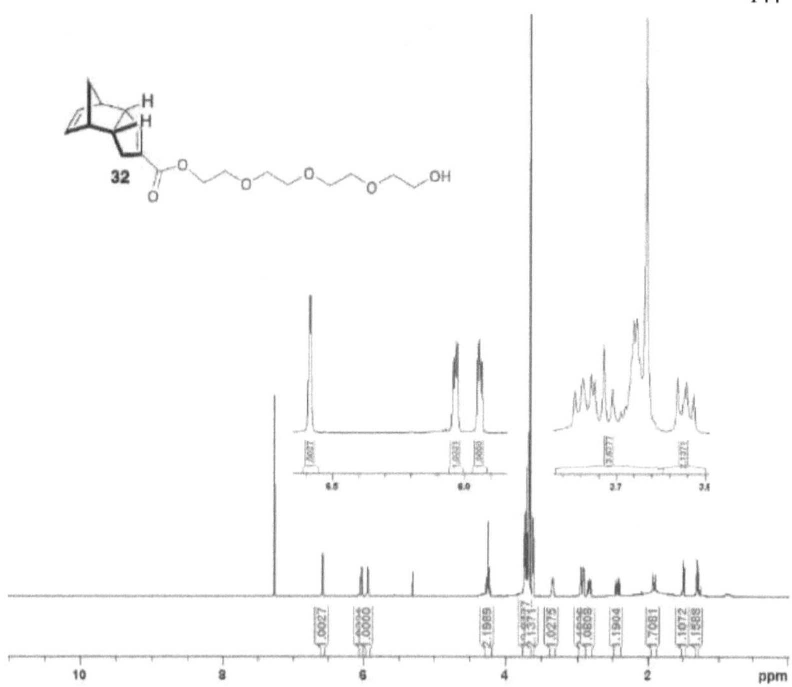

Figure A 29. ^{1}H NMR spectrum for *f*DCPD–TEG ester in CDCl$_3$, recorded at 500 MHz. The inset spectra show expansions of key regions of the spectrum.

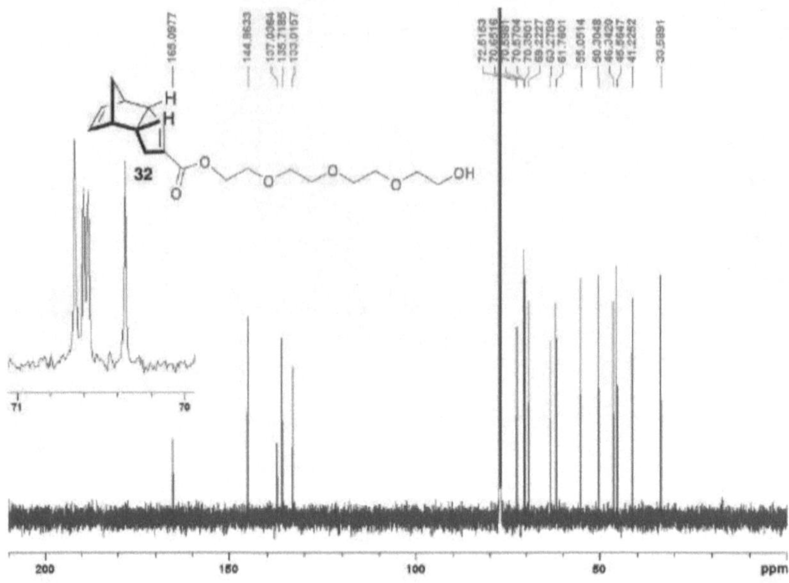

Figure A 30. ^{13}C NMR spectrum for *f*DCPD–TEG esters in CDCl$_3$, recorded at 125 MHz. The inset spectrum shows an expansion in the region between 70-71 ppm, to highlight the distinct resonances visible in this region.

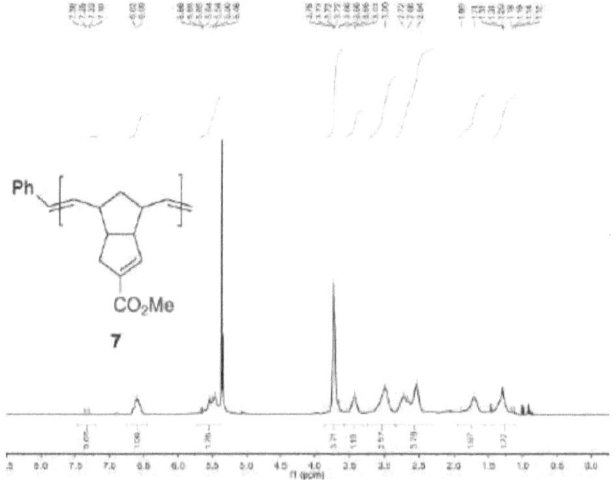

Figure A 31. ^{1}H NMR spectrum of polymer 7 (300 MHz, CD$_2$Cl$_2$).

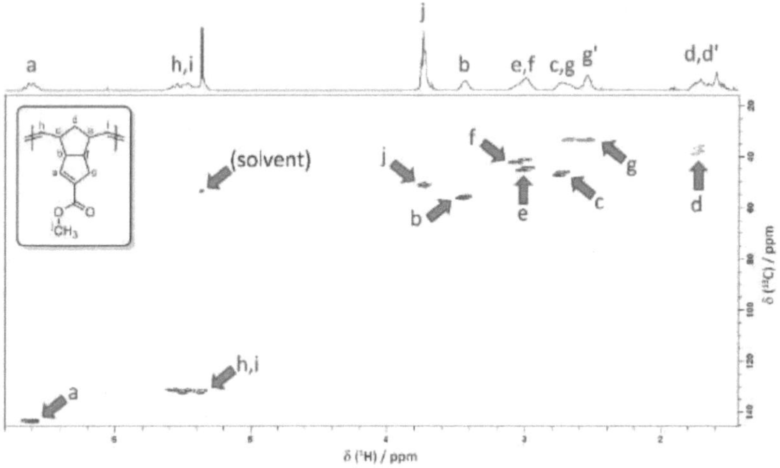

Figure A 32. Full-range HSQC spectrum for linear *f*PDCPD polymer 7, produced from column-purified monomer 4, in CD$_2$Cl$_2$.

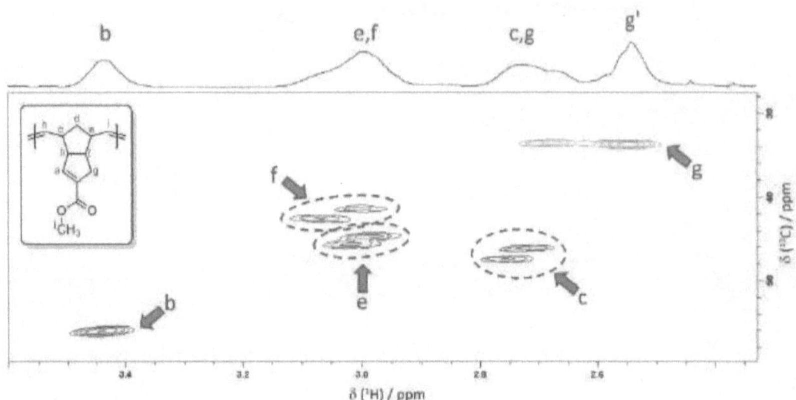

Figure A 33. Close-up of HSQC spectrum for 7, emphasizing the duplication of ^{13}C signals for some carbon atoms. This duplication could be due to the use of racemic monomer 4, (which necessarily leads to the production of atactic polymer) but more likely is due to the possibility of both head→head and head→tail connectivity along the backbone (i.e. the primary polymer linkage could be h,c,d,e,i→h,c,d,e,i or h,c,d,e,i→i,e,d,c,h at any given backbone alkene).

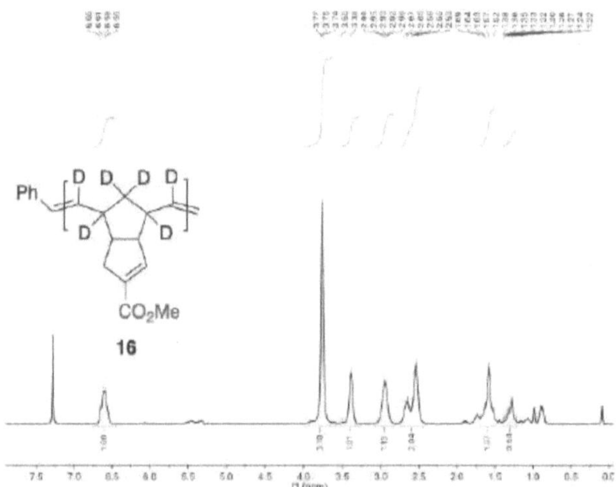

Figure A 34. ^{1}H NMR spectrum of polymer 16 (300 MHz, CDCl$_3$).

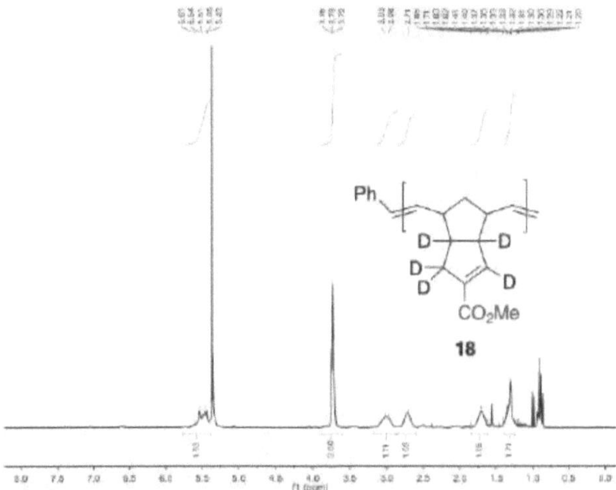

Figure A 35. ^{1}H NMR spectrum of polymer 18 (300 MHz, CD$_2$Cl$_2$).

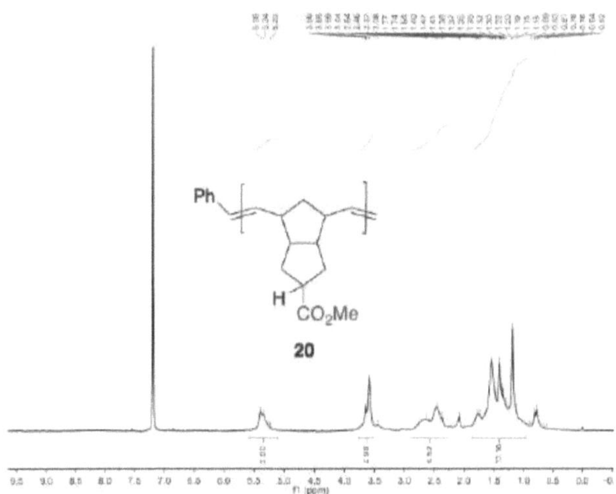

Figure A 36. ^{1}H NMR spectrum comparison of polymer 7 and 20 (300 MHz, CDCl$_3$).

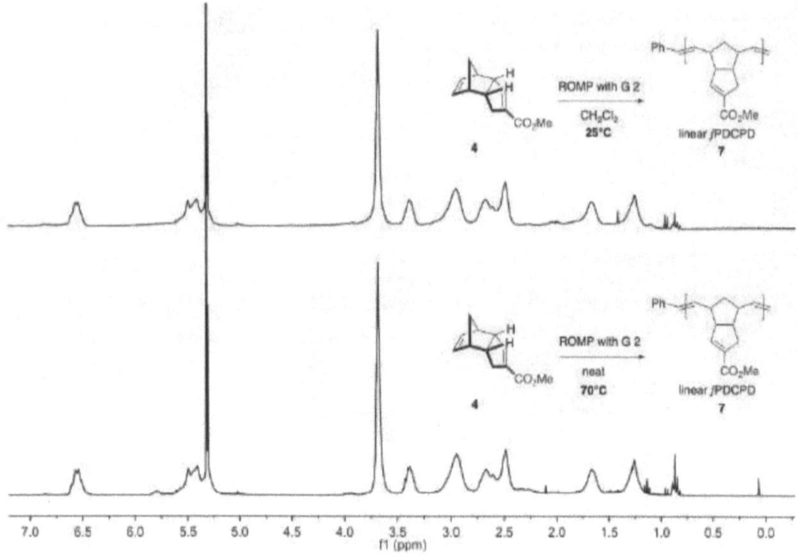

Figure A 37. Comparison of ^{1}H NMR data for linear polymer 7 prepared in solution at room temperature, or neat at 70°C.

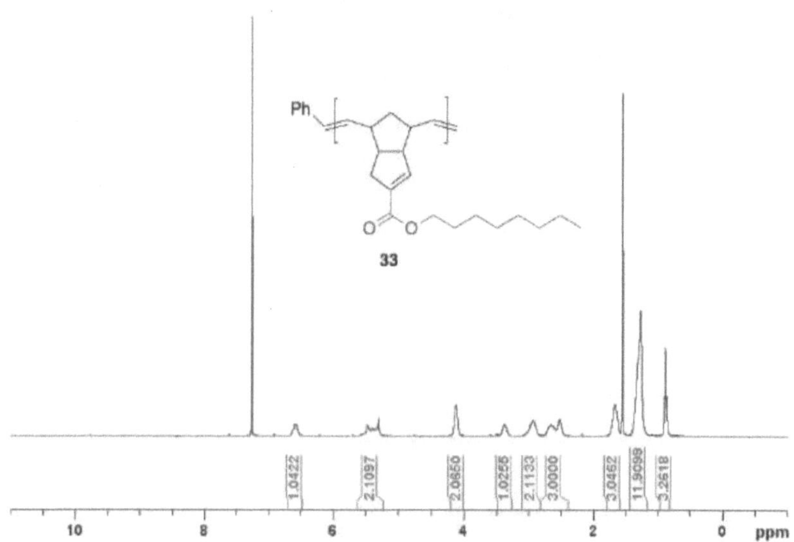

Figure A 38. ^{1}H NMR spectrum for ƒPDCPD-octyl ester **33** (linear precursor to crosslinked polymer **27** in CDCl$_3$, recorded at 300 MHz.

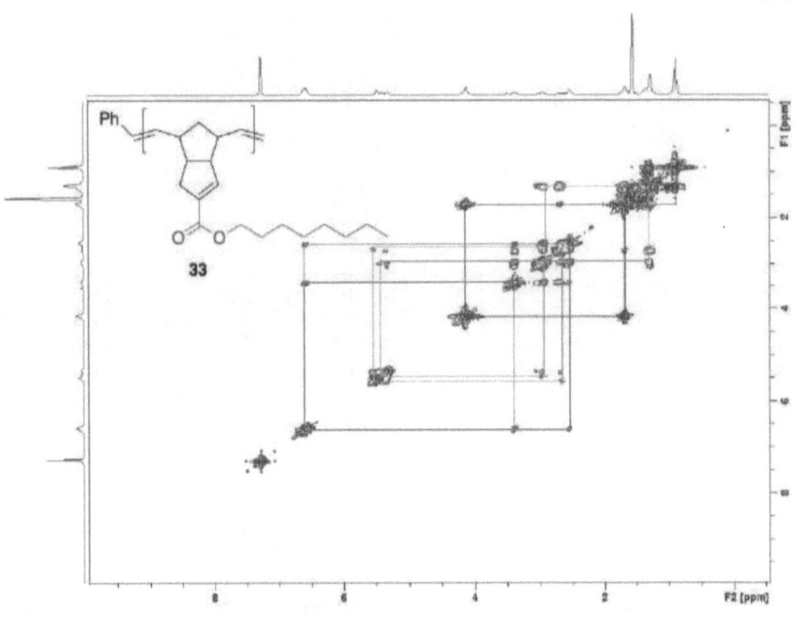

Figure A 39. 500 MHz COSY NMR spectrum for *f*PDCPD-octyl ester **33** (linear precursor to crosslinked polymer **27**) in CDCl₃. Colored boxes indicate correlations between ¹H signals.

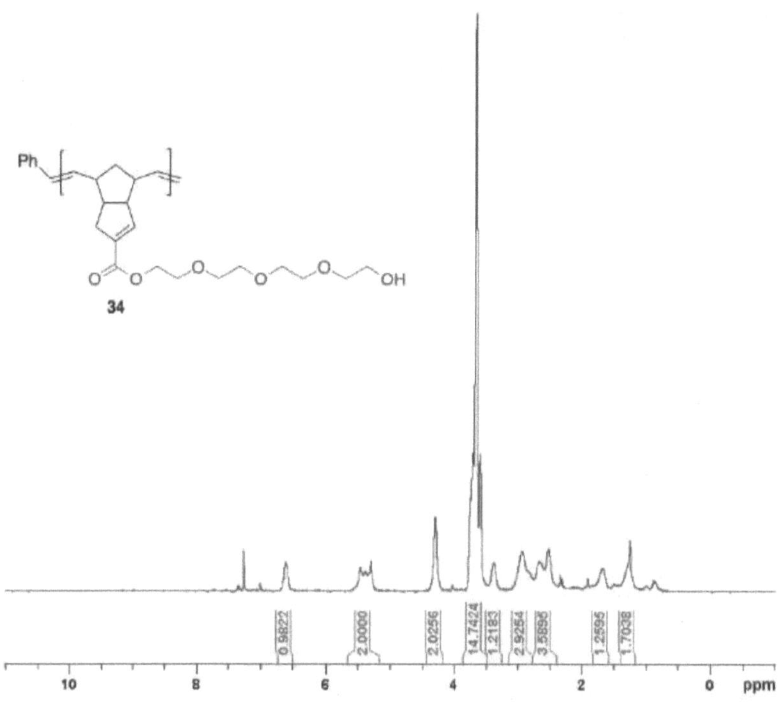

Figure A 40. ¹H NMR spectrum for *f*DCPD–TEG ester (linear precursor to crosslinked polymer 28) in CDCl₃, recorded at 300 MHz.

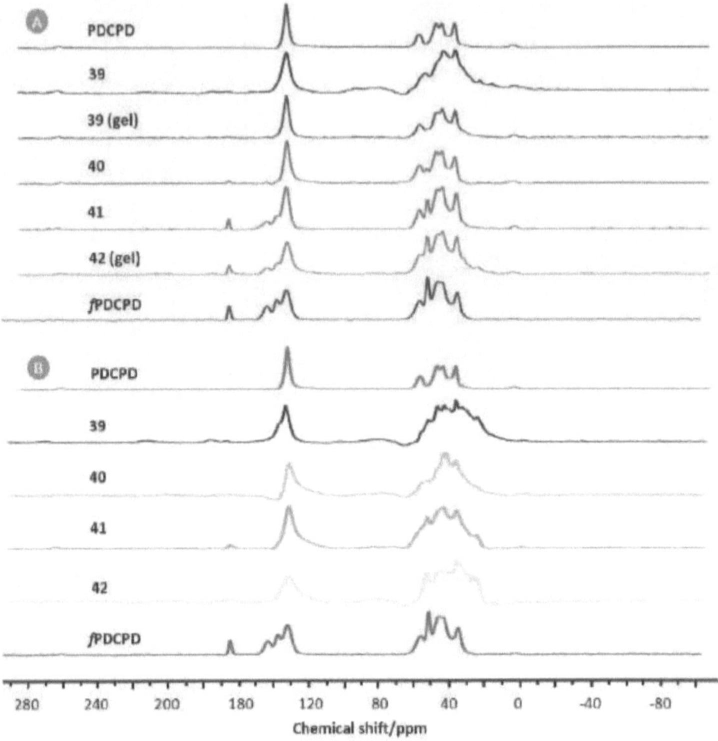

Figure A 41. Solid-state ^{13}C{^{1}H} NMR spectrum of A) first and B) second batch of synthesized homo- and copolymers prior to thermal curing. Data shown in Figure 30 in chapter 5 are from the third batch of synthesized material.

Appendix B: FTIR Spectrum for Selected Compounds

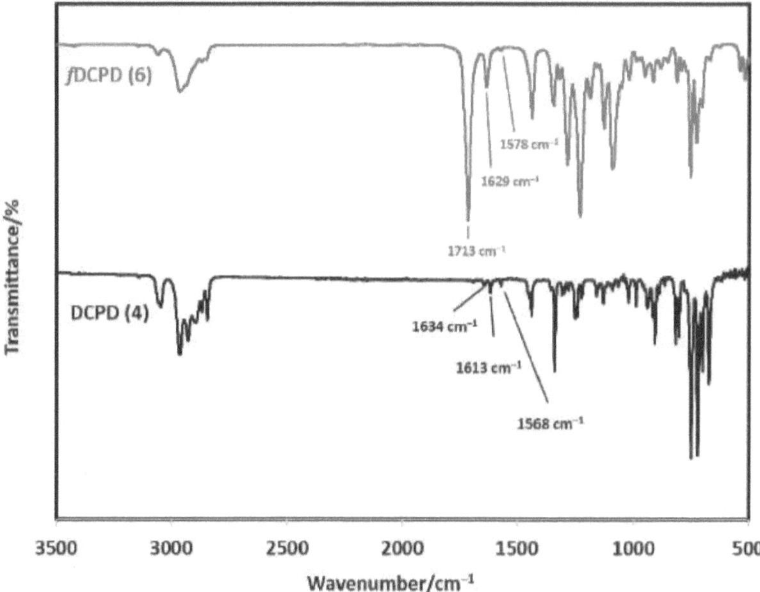

Figure B 1. FTIR spectra of DCPD and *f*DCPD monomers.

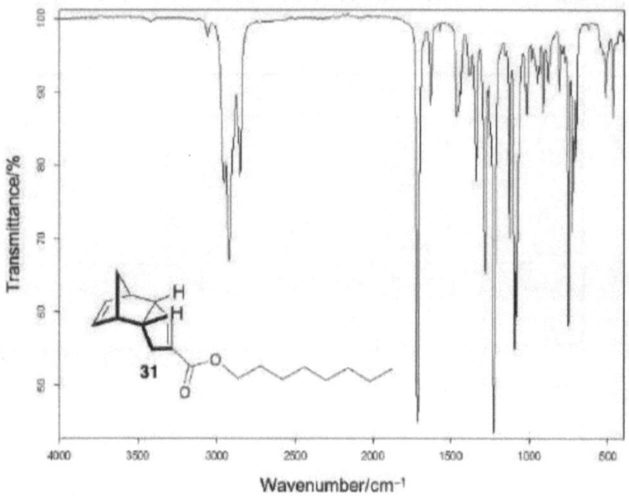

Figure B 2. FTIR spectrum for octyl–ester functionalized dicyclopentadiene 31.

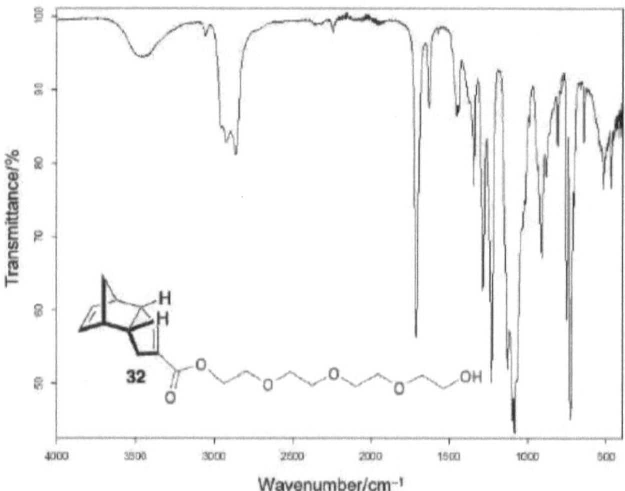

Figure B 3. FTIR spectrum for TEG–ester functionalized dicyclopentadiene 32.

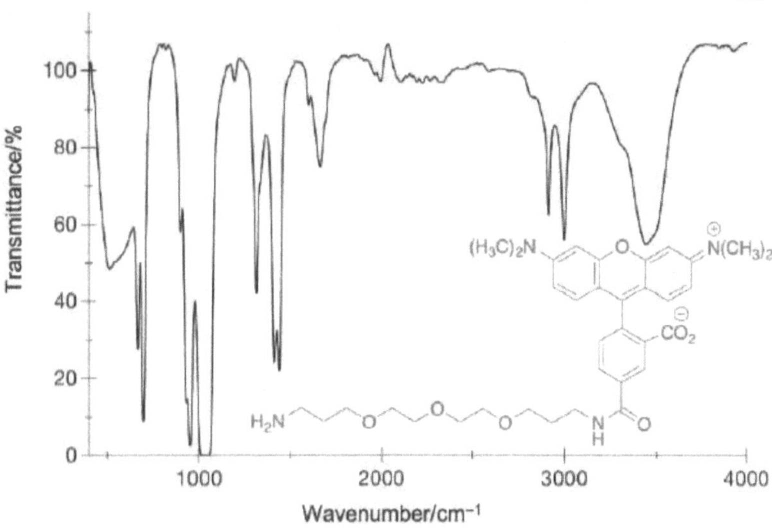

Figure B 4. FTIR spectrum for 5-Tamra-PEO3-amine in DMSO.

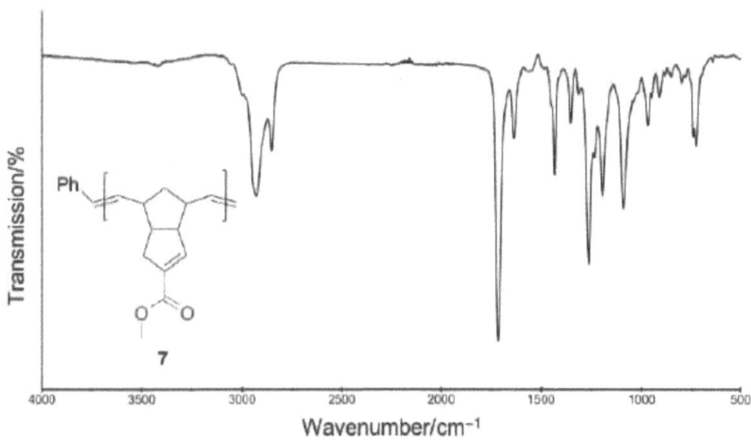

Figure B 5.FTIR spectrum of polymer 7.

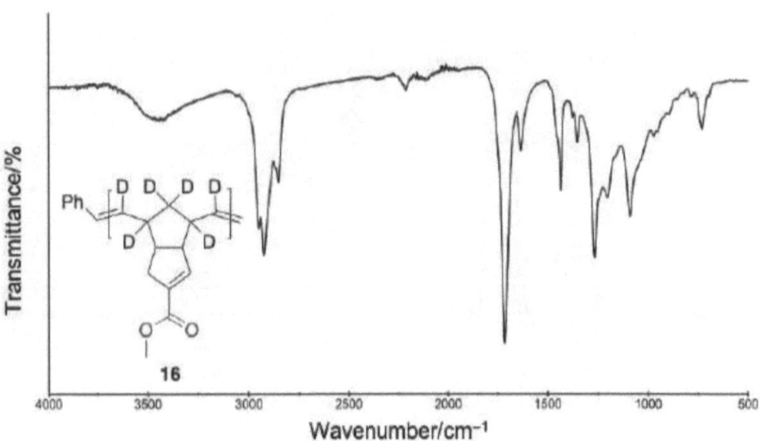

Figure B 6. FTIR spectrum of half-deuterated ƒPDCPD 16.

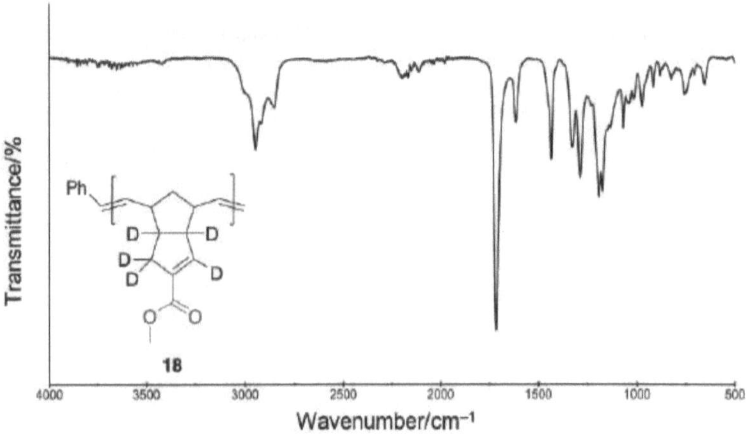

Figure B 7. FTIR spectrum of of half-deuterated ƒPDCPD 18.

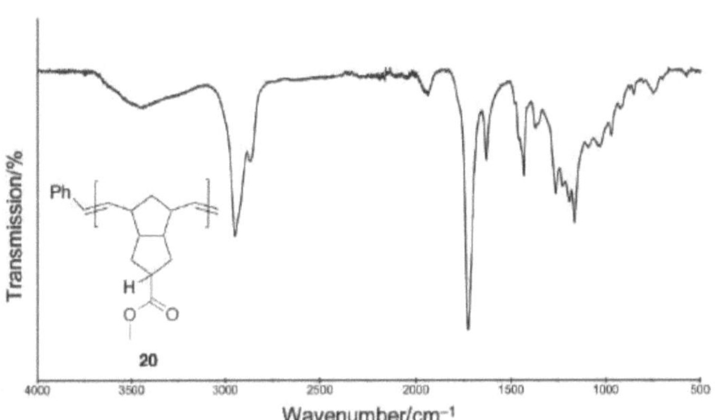

Figure B 8. FTIR spectrum of hydrogeneated *f*PDCPD 20.

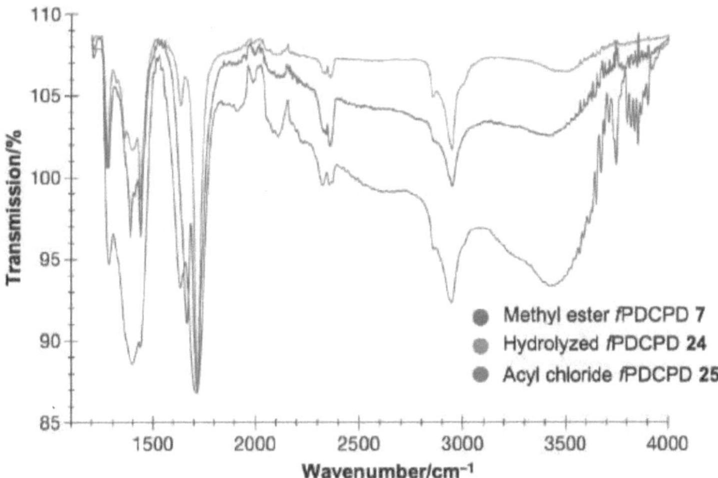

Figure B 9. ATR-FTIR spectra for methyl ester *f*PDCPD 7, partially hydrolyzed *f*PDCPD 24 and the putative acyl-chloride-containing *f*PDCPD 25 on glass cover slides. The characteristic acyl chloride carbonyl stretch is not visible in this sample, either due to decomposition during analysis or else due to constraints of the surface-bound sample. The corresponding small-molecule analogue, *f*DCPD–COCl (compound 30), formed under identical conditions, was confirmed by a downfield shift of the vinyl proton in the ¹H NMR spectrum Figure A26.

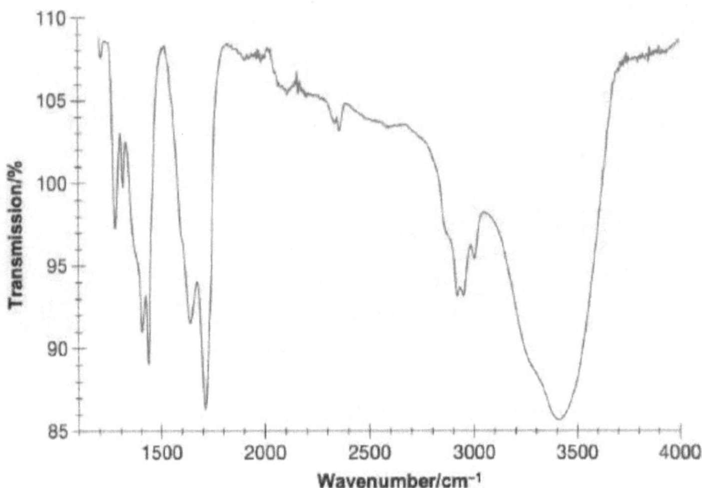

Figure B 10. FTIR spectrum of *f*DCPD spin-coated cover slides functionalized with TAMRA (compound 26).

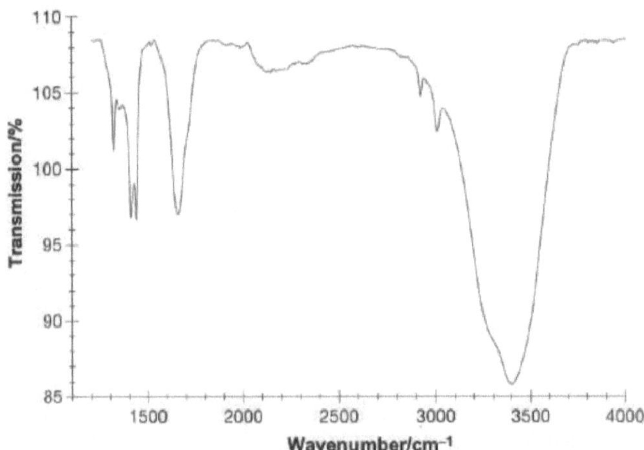

Figure B 11. FTIR spectrum for *f*PDCPD spin-coated cover slides functionalized with chloramphenicol (compound 36).

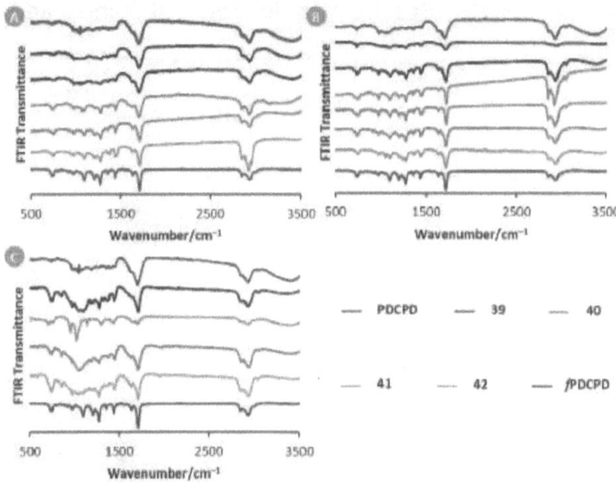

Figure B 12. FTIR spectra for A) first batch B) second batch and C) third batch of synthesized homo polymers and copolymers prior to thermal curing. Data shown in Figure 31 in chapter 5 are from the third batch of synthesized material.

Appendix C: Raman Spectral Data for Selected Compounds

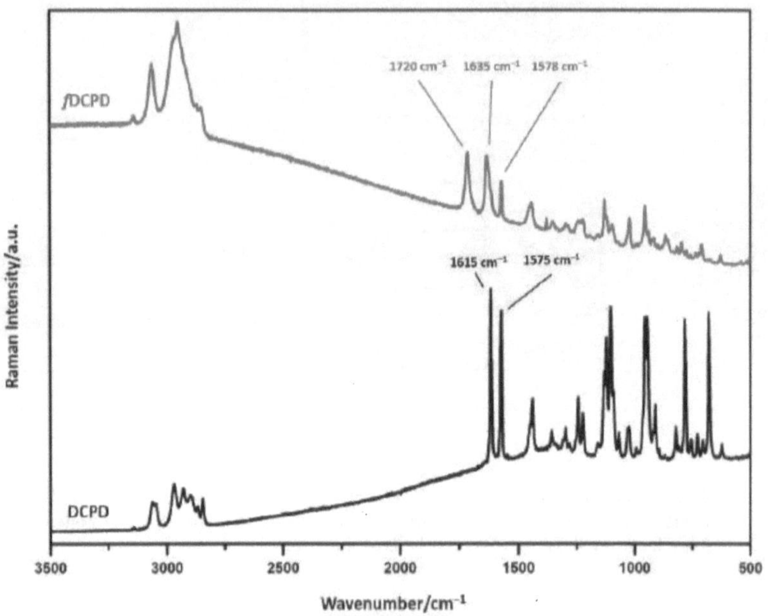

Figure C 1. Raman spectra of DCPD and *f*DCPD monomers.

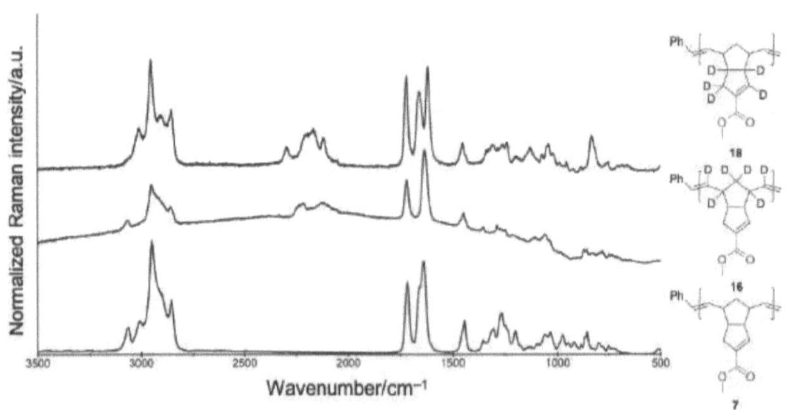

Figure C 2. Raman spectra of 7 (bottom), 16 (middle), and 18 (top).

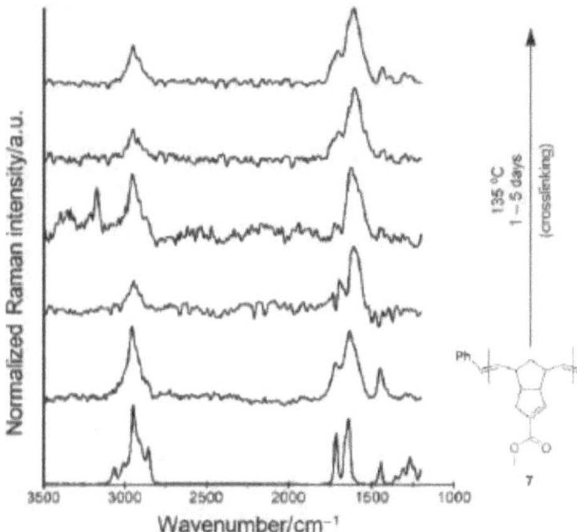

Figure C 3. Changes in Raman spectra following crosslinking of the parent, undeuterated polymer 7.

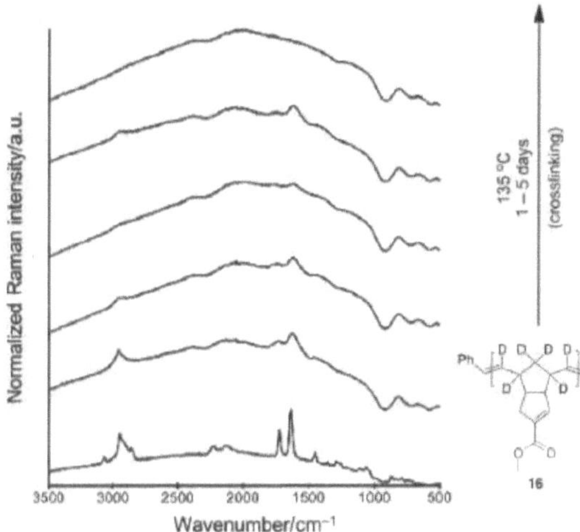

Figure C 4. Changes in Raman spectra following crosslinking of the deuterated polymer 16.

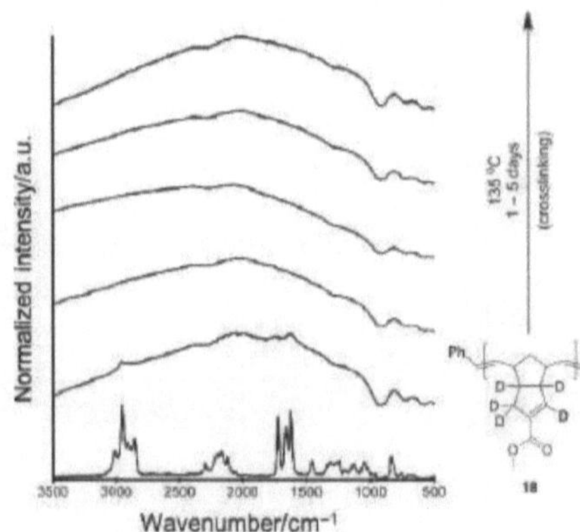

Figure C 5. Changes in Raman spectra following crosslinking of the deuterated polymer 18.

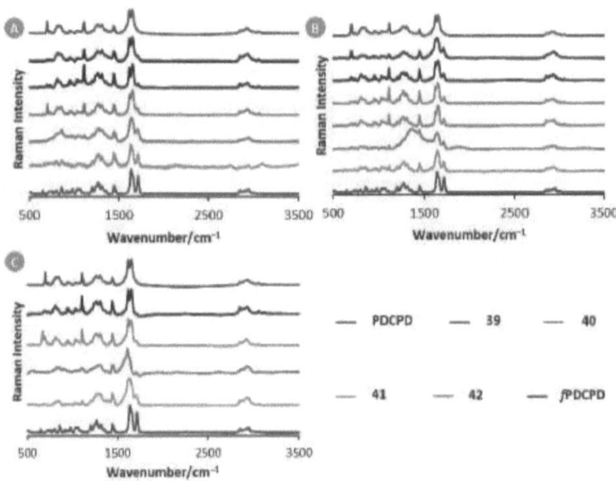

Figure C 6. Raman spectra for A) first batch B) second batch and C) third batch synthesized homo polymers and copolymers prior to thermal curing. Data shown in Figure 31 in chapter 5 are from the third batch of synthesized material.

Appendix D: Thermomechanical Testing Results

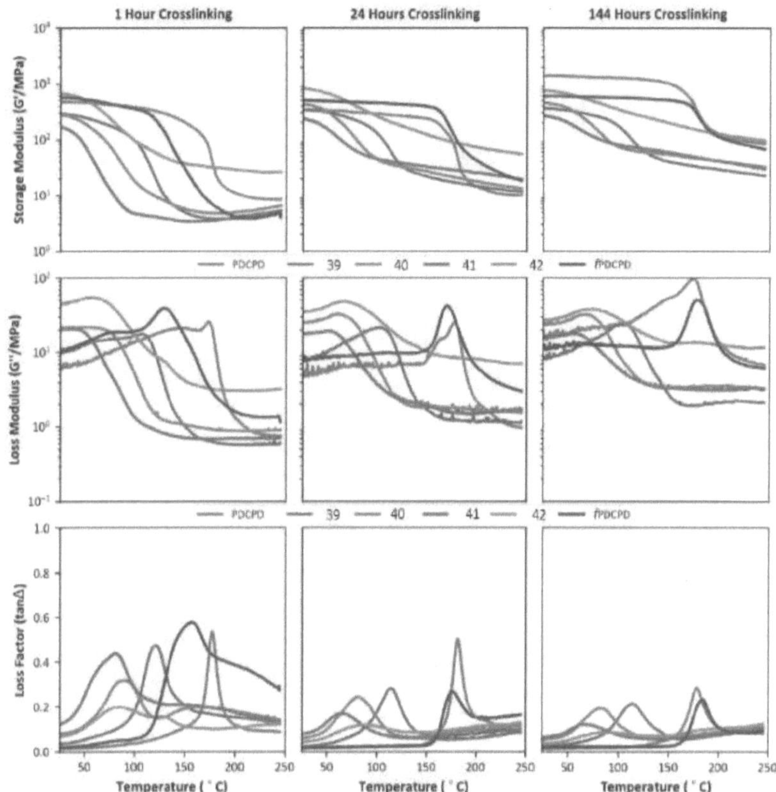

Figure D 1. DMA plots of first batch of synthesized homo- and copolymers and copolymers after thermal crosslinking. Plots were generated by Mr. Canyu (Charles) Cai using python scripts.

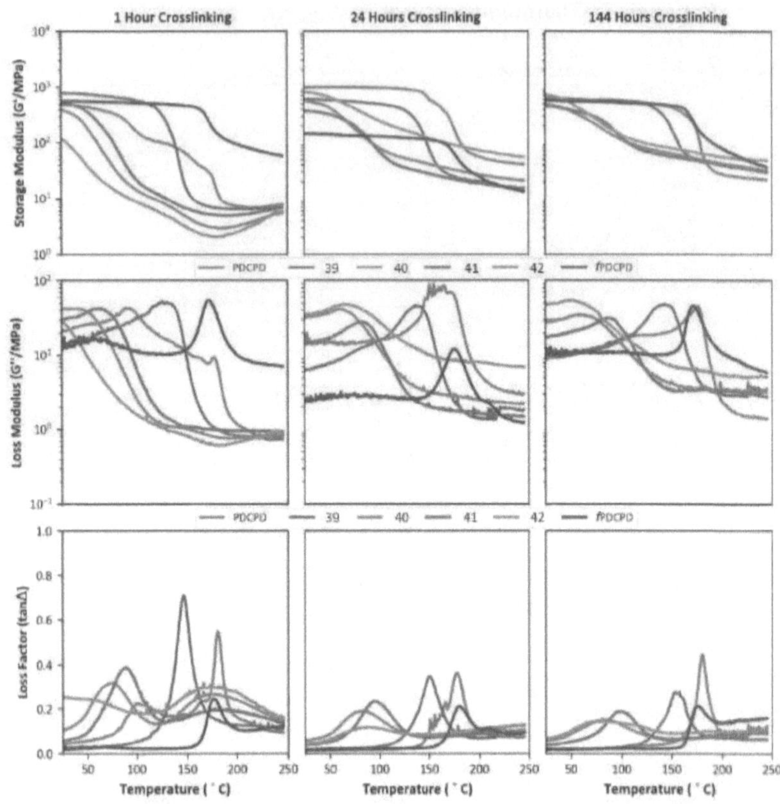

Figure D 2. DMA plots of second batch of synthesized homo- and copolymers and copolymers after thermal crosslinking. Plots were generated by Mr. Canyu (Charles) Cai using python scripts.

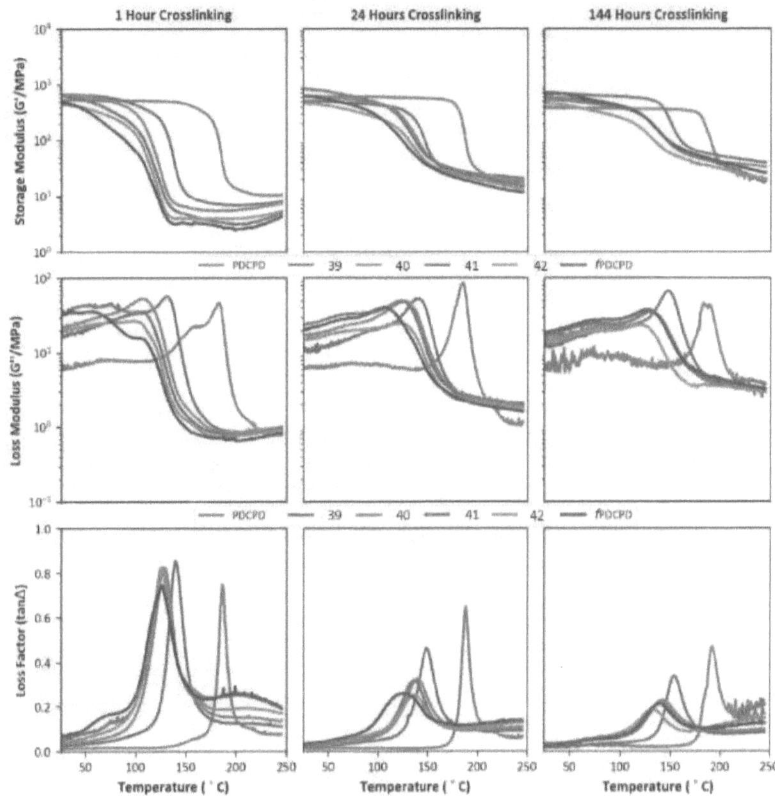

Figure D 3. DMA plots of third batch of synthesized homo- and copolymers and copolymers after thermal crosslinking. Plots were generated by Mr. Canyu (Charles) Cai using python scripts.

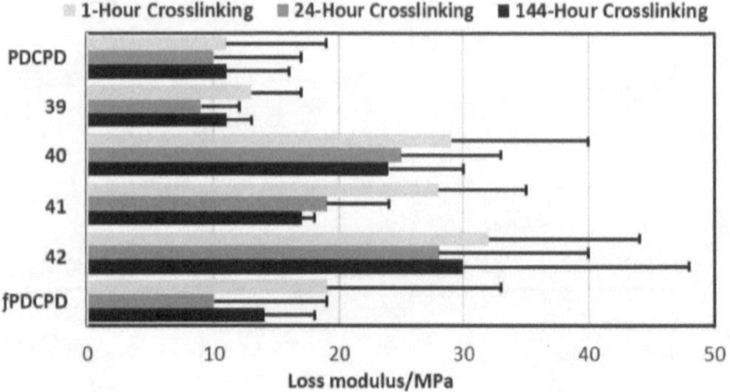

Figure D 4. Comparison of loss modulus at room temperature (25 °C) for synthesized homopolymers and copolymers after thermal curing. Samples were analyzed in triplicate, using three different preparations. Error bars represent standard deviation.

Table D 1. Alpha transition temperature for homopolymers and copolymers after thermal crosslinking. Samples were analyzed in triplicate, using three different preparations. SD represents standard deviation.

Temperature (°C)	1-Hour Crosslinking	SD	24-Hour Crosslinking	SD	144-Hour Crosslinking	SD
ƒPDCPD	153	25	160	30	166	24
42	128	42	101	22	99	29
41	102	30	96	32	102	37
40	102	34	99	26	101	36
39	139	15	135	19	142	24
PDCPD	182	6	182	4	183	8

Table D 2. Storage modulus at room temperature (25 °C) for homo polymers and copolymers after thermal crosslinking. Samples were analyzed in triplicate, using three different preparations. Error bars represent standard deviation.

G' (MPa)	1-Hour Crosslinking	SD	24-Hour Crosslinking	SD	144-Hour Crosslinking	SD
ƒPDCPD	538	36	419	247	629	78
42	409	279	695	206	648	161
41	440	242	479	312	434	157
40	442	196	537	105	486	39
39	573	255	482	136	549	169
PDCPD	497	31	640	322	790	526

Table D 3. Loss modulus at room temperature (25 °C) for homopolymers and copolymers after thermal crosslinking. Samples were analyzed in triplicate, using three different preparations. Error bars represent standard deviation.

G'' (MPa)	1-Hour Crosslinking	SD	24-Hour Crosslinking	SD	144-Hour Crosslinking	SD
ƒPDCPD	19	14	10	9	14	4
42	32	12	28	12	30	18
41	28	7	19	5	17	1
40	29	11	25	8	24	6
39	13	4	9	3	11	2
PDCPD	11	8	10	7	11	5

Table D 4. Vickers hardness number (HV) and converted modulus number for homopolymers and copolymers after thermal crosslinking. Samples were analyzed in triplicate, using three different preparations. Error bars represent standard deviation.

Vickers hardness Number (HV)				
	1 day	SD	6 days	SD
ƒPDCPD	19.5	1.45	24.1	1.18
42	12.3	4.85	18.4	4.48
41	14.5	6.75	18.3	3.44
40	11.3	5.88	15.6	4.94
39	11.3	4.11	15.5	2.9
PDCPD	13.4	2.56	14.8	0.75
Convert to Modulus (MPa)				
	1 day	SD	6 days	SD
ƒPDCPD	191.2	14.22	236.3	11.57
42	120.6	47.56	180.4	43.94
41	142.2	66.2	179.5	33.74
40	110.8	57.67	153	48.45
39	110.8	40.31	152	28.44
PDCPD	131.4	25.11	145.1	7.36

Equation D 1. Vickers hardness number convert to modulus.

$$HV = 0.1891 \frac{F}{d^2} \left[\frac{N}{mm^2} \right]$$

Where HV is Vickers hardness number converted to modulus, F is the force applied to sample and d is the diameter of the cylinder of samples. To simplify the equation: $Modulus = HV * g$. To convert HV to MPa, $g = 9.807$. Equations were adapted from
i) https://www.gordonengland.co.uk/hardness/hvconv.htm
Ii) https://www.gordonengland.co.uk/hardness/vickers.htm.

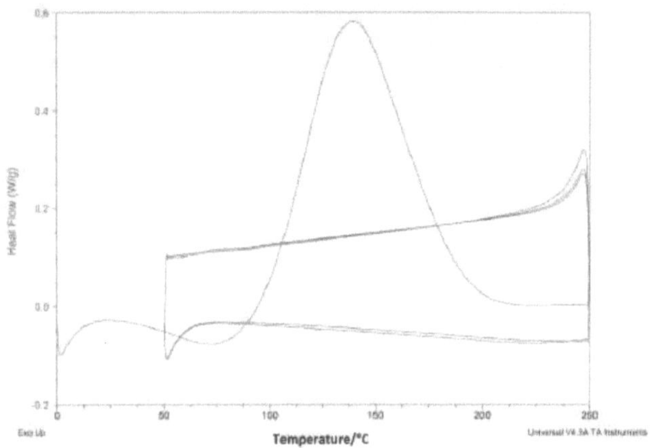

Figure D 5. Representative DSC example for copolymers 39. No clearly defined glass transition temperature was observed.

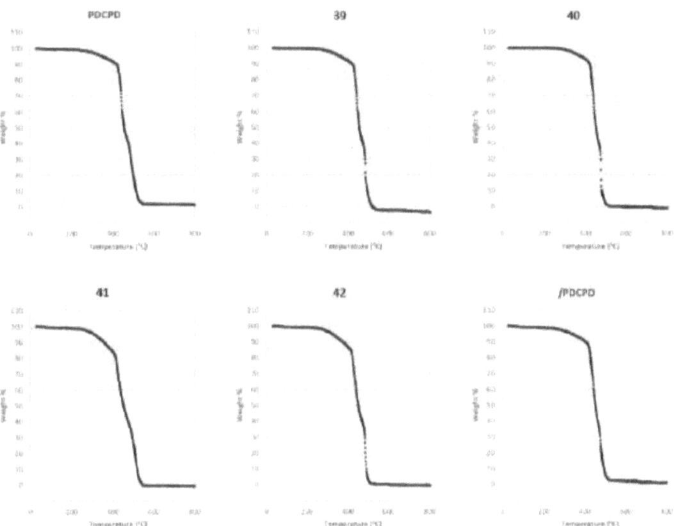

Figure D 6. Collected TGA data for all 6 polymers and copolymers, following thermal curing.

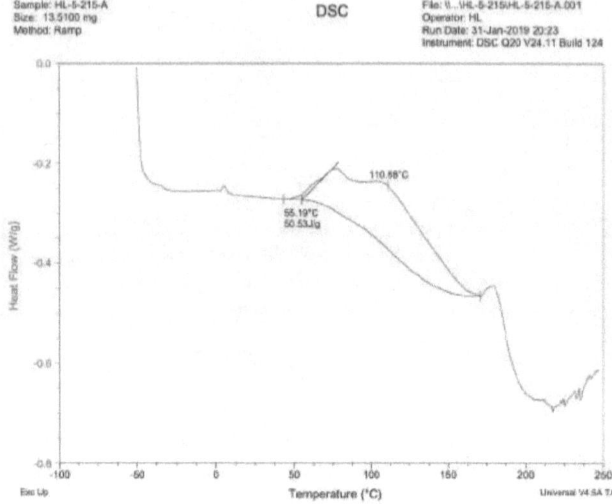

Figure D 7. Thermal investigation of *f*DCPD polymerization by DSC. ΔH is not high enough to maintain a FROMP.

Appendix E: Other Chemical Structures and Kinetic Study of Polymerization

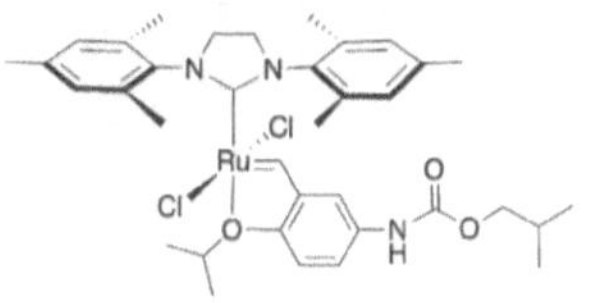

Figure E 1. Chemical structure of Umicore M73 catalyst.

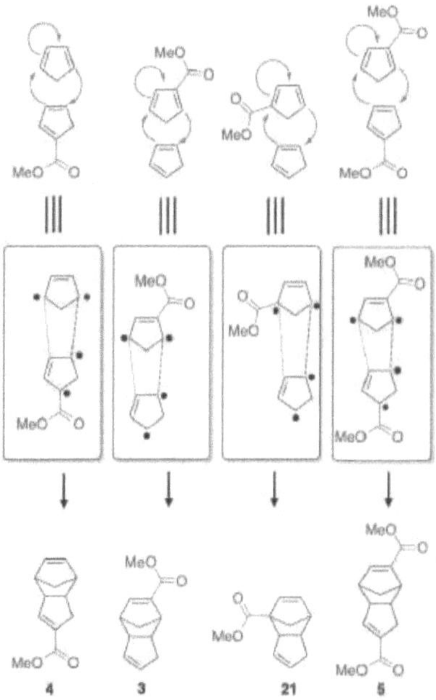

Figure E 2. The regiochemical outcome for the Diels–Alder reactions leading to each of the ester-functionalized dicyclopentadienes **4**, **3**, **21**, and **5** (as well as the unfunctionalized dicyclopentadiene **6**, not shown), is consistent with our earlier prediction of regiochemistry using radical stabilization arguments.[40,41] Briefly, these arguments require that each diene species be viewed as its corresponding 1,4-diradical resonance structure. The fastest Diels–Alder coupling is then predicted to arise through the coupling of the least-stabilized radicals (highlighted in yellow in the Figure). The characterization of the new compound **21** (which was not known at the time of our earlier publications) therefore further supports these arguments.

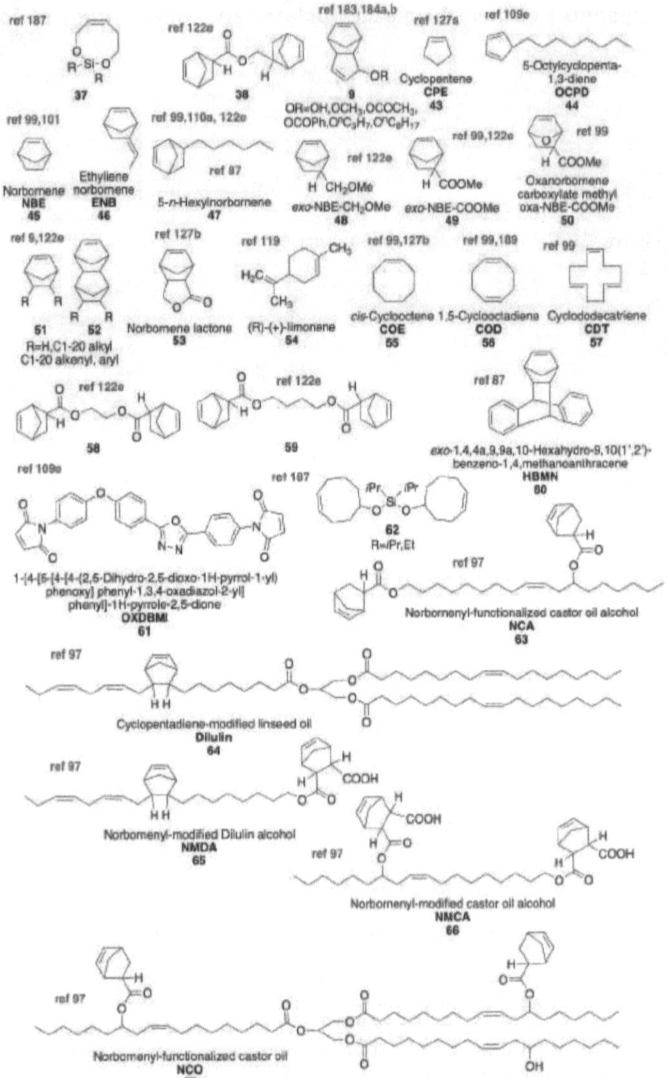

Figure E 3. Full list of representative small molecules that have been copolymerized with DCPD or used in related experiments.

Table E 1. Increase in molecular weight with decreasing concentration.

Catalyst loading	Integration		Repeat units	MW
(mol%)	vinyl H	phenyl H	(n)	(g/mol)
10	1	0.1547	32.3	6147
5	1	0.135	37	7044
2.5	1	0.1264	39.6	7524
1	1	< 0.08	> 60	> 10000
0.1	1	< 0.02	> 200	> 40000

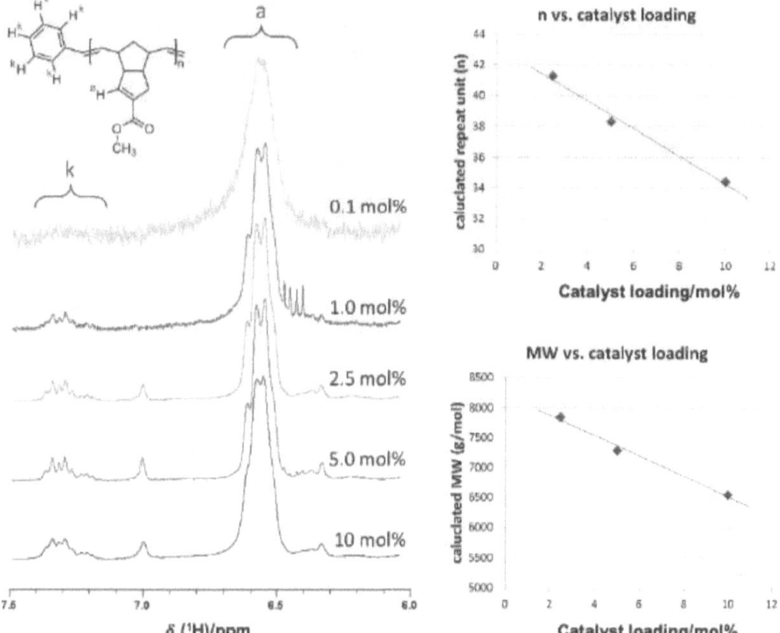

Figure E 4. Kinetic study of ring-opening metathesis polymerization Increase in molecular weight with decreasing catalyst concentration.

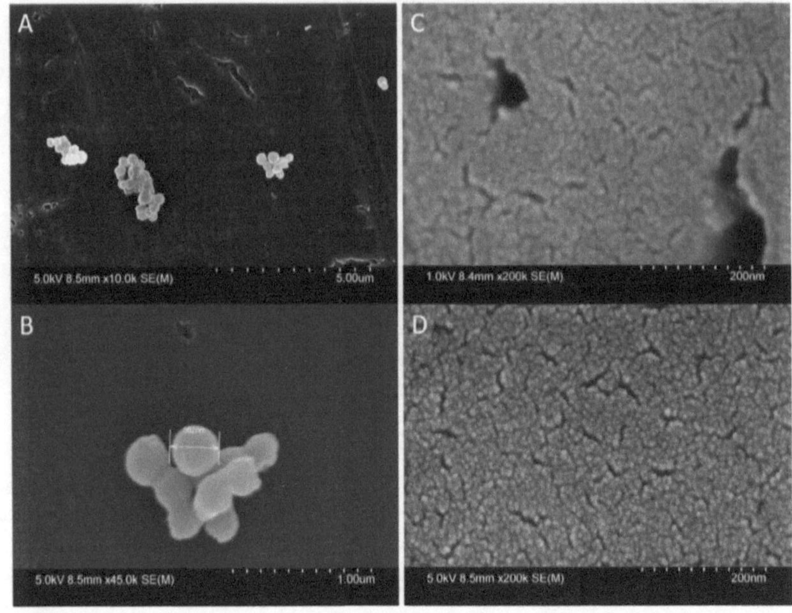

Figure E 5. SEM data for copolymer 39. The ca. 350 nm particles visible on top of the polymer surface in panels A and B are artifacts of the process used to gold-coat the sample.

Appendix F: Photos of "UVIC" letters

Figure F1. UVIC letters. Photo taken by Ella (Yu) Guan.